Ideas, Machines, and Values:

An Introduction to Science, Technology, and Society Studies

Stephen H. Cutcliffe

ROWMAN & LITTLEFIELD PUBLISHERS, INC.
Lanham • Boulder • New York • Oxford

ROWMAN & LITTLEFIELD PUBLISHERS, INC.

Published in the United States of America
by Rowman & Littlefield Publishers, Inc.
4720 Boston Way, Lanham, Maryland 20706
http://www.rowmanlittlefield.com

12 Hid's Copse Road
Cumnor Hill, Oxford OX2 9JJ, England

British Library Cataloguing in Publication Information Available

Library of Congress Cataloging-in-Publication Data
Cutcliffe, Stephen H.
Ideas, machines, and values : an introduction to science, technology, and society studies / Stephen H. Cutcliffe.
p. cm.
Includes bibliographical references and index.
ISBN 0-7425-0066-7 (cloth : alk. paper)—ISBN 0-7425-0067-5 (pbk. : alk. paper)
1. Science—Social aspects. 2. Science—Philosophy. 3. Technology—Social aspects. 4. Technology—Philosophy. I. Title.

Q175.5.C88 2000
303.48'3--dc21

99-057149

Printed in the United States of America

The paper used in this publication meets the minimum requirements of American National Standard for Information Sciences—Permanence of Paper for Printed Library Materials, ANSI/NISO Z39.48–1992.

To Barbara, Nicolas, and Daniel,
who gave me the time and space

Contents

Preface and Acknowledgments

The locomotive and the steamboat, like enormous shuttles shoot every day across the thousand various threads of national descent and employment and bind them fast in one web.
— Ralph Waldo Emerson, "The Young American," 1844

When we try to pick out anything by itself, we find it hitched to everything else in the universe.
— John Muir, *My First Summer in the Sierra*, 1911

Thoughtful people have long recognized the societal significance of science and technology. Thus it should come as no surprise to find the insightful transcendentalist Ralph Waldo Emerson commenting in his 1844 essay, "The Young American," on the societal interconnectedness of the then widely emerging transportation technologies of steamboats and railroads. Nor should we be astonished at the early environmental leader John Muir's recognition of the interrelationships among the things of life. Both men had intuitively grasped the interconnectedness of life, especially with regard to the relationships among science, technology, and society—among ideas, machines, and values.[1] The adoption, especially by Emerson, of a weaving metaphor implies the early recognition of the "seamless web," as contemporary sociologists and historians are wont to put it, characterizing the holistic interdependencies—the warp and the woof—of science and technology on the one hand, and of society on the other, in modern America. Given that early recognition, it is rather surprising that it took more than a century after Emerson for anything approaching a formal interdisciplinary study of science and technology in their societal context to emerge. I should note here that while social studies of medicine should really be included alongside those of science and technology, medicine generally has not been a central feature, being left to develop in the main as a separate field. While there were certainly disciplinary efforts to study the history, the philosophy, and later the sociology of science, technology, and medicine dating from the first part of the twentieth century, it was not until roughly the mid- to late 1960s that what has become known as the field of Science, Technology, and Society

(STS) studies formally developed. This book is intended as an introductory overview of the emergence of STS as a field of study, and as a summary of its current interests and concerns, with some suggestions regarding its future development.

STS, or Science and Technology Studies as it is sometimes now called, has as its primary focus the explication and analysis of science and technology as complex societal constructs with attendant societal influences entailing a host of epistemological, political, and ethical questions. In this "contextual" view, STS has come to recognize science and technology as neither wholly autonomous juggernauts nor simply neutral tools ready for any utilization whatsoever. Instead, science and technology are commonly perceived as value-laden social processes taking place in specific historical contexts shaped by, and in turn shaping, the human values reflected in cultural, political, and economic institutions. Such a view does not deny the constraints imposed by nature or the physical reality of technological artifacts, but it does insist that our knowledge and understanding of nature, of science, and of technology are socially mediated processes. It is toward just such a holistic and interdisciplinary understanding, coupled with the hope that society will be better able to shape and control its science and technology as a result, that STS aims and is particularly well suited to assist us.

As a field of academic study, and even as a social movement, STS proposes neither unbridled enthusiasm for, nor absolute rejection of, science and technology. Rather it seeks to walk the line of judicious analysis and thoughtful application between the poles of untempered optimism and nihilistic pessimism. We live in an age in which there are a great many things to appreciate about contemporary science and technology in terms of their contribution to our knowledge bases and to the enhancement of our daily lives. Yet we must also be equally cognizant of some of the negative consequences that almost inevitably follow in their wake, at least for some people. At the turn of the millennium, society is faced with both the promises and the dangers of scientific and technological endeavors such as the human genome project and electronic communication systems, developments that must surely change our lives, either for better or for worse, but most likely for both. STS can contribute to our understanding of such issues by serving as a dialectical synthesis between the optimistic "Better Living through Chemistry" and its neo-Luddite antithesis. It can enhance our scientific and technological literacies, including the inherent societal and political contexts, so that we might better understand and deal with such issues as global climate change, no matter what our particular perspective on the most appropriate paths to follow. In sum, STS offers society a window through which to view reflexively its own interactions with science and technology. At the same time, this window provides a framework by which to structure a more democratic societal control over contemporary technoscience.

In contemporary high-tech society, it is often difficult, if not impossible, to draw clear-cut lines between that which constitutes science and that which constitutes technology, and in some ways even from that which constitutes society. The French sociologist of science Bruno Latour and some other European STS scholars have tended to adopt the term "technoscience" to emphasize the inseparability of these phenomena. As Latour argued in a 1998 essay in *Science* on the occasion of the American Association for the Advancement of Science's 150th anniversary, "science and society are now entangled to the point where they cannot be separated any longer." In another *Science* anniversary essay, Tsou Chen-Lu, director of the Department of Life Sciences at the Chinese Academy of Sciences in Beijing, noted that today "the link between science and technology is so intimate that they have merged into one term in the Chinese language, 'keji,' which means scitech." In recognition of such interlinkages, I too have adopted the term "technoscience" throughout much of this book when referring to the dynamic interrelationships between science and technology. I also find it useful as a shorthand replacement for the cumbersome and oft repeated phrase "science and technology."[2]

I first became interested in STS as a field of study in the mid-1970s when I began working with what was then called the Humanities Perspectives on Technology Program at Lehigh University. In 1977 the HPT Program changed its name to Science, Technology, and Society. As editor of the *Science, Technology & Society Curriculum Newsletter* for more than two decades I have been witness, almost from the beginning, to many of the developments within STS, especially from the perspective of curricular developments. Many aspects of the field have changed during this period, and what follows is my attempt to make some sense out of that history. Chapter 1 offers a brief historical overview of the development of STS as a social movement and as an academic field of study, from its inception in the mid-1960s. As such it sets up the more in-depth analyses in the chapters that follow. Chapter 2 provides a summary of key disciplinary developments in those fields central to the development of STS through roughly the mid-1980s: history, sociology, and philosophy of science and of technology. Chapter 3 picks up the chronological story at the point where STS may be said to have become more interdisciplinary in nature. It does so first through a discussion of the issue of interdisciplinarity itself and then shifts to a discussion of where STS scholarship is today. In particular it looks at the influence of cultural studies and the resultant "science wars" phenomenon. Chapter 4 takes an institutional turn away from the focus on scholarly literature in order to describe STS programs and organizations that have evolved to help establish its boundaries. Finally, chapter 5 asks the questions, "Why should we study STS?" and "In what directions should STS be headed?" Following a brief set of conclusions, I have also included a bibliographical essay identifying important works in the STS field, as well as a more selective listing of one hundred key works.

Throughout the book I have drawn heavily on experiences and developments in North America, especially the United States, which is what I have experienced directly; however, when and where possible I have included material on developments in Western Europe and in other parts of the world such as China and Australia. The resulting overview will, I hope, be of interest to beginning students of STS at both the undergraduate and graduate levels. I hope as well that the book will serve scholars, especially those trained in fairly traditional disciplines, who may be venturing into the more interdisciplinary world of STS for the first time and who may need at least a broad-scale map by which to navigate.

The original genesis of this volume was a request for an essay summarizing the emergence of STS as a field of study from friend and professional colleague, Carl Mitcham, who was then editing a special issue of *Research in Philosophy and Technology*. It is to him that I owe a primary debt of gratitude for initiating this project, and more importantly for encouraging me, sometimes with a sharp prod, to expand my ideas into what has ultimately become this volume. Since that time I have had the opportunity to extend my thinking on some of those early ideas through invitations to lecture and publish on the topic of the STS field's development. I thus also owe thanks to Sheila Jasanoff, Cornell University; John Hunter, Alfred State College; Manuel Medina, University of Barcelona; José Sanmartín, University of Valencia; Yin Deng-xiang, Chinese Academy of Social Sciences; and Li Bocong, Chinese Academy of Science for invitations to speak at conferences where I was able to elaborate further on my perceptions.

Several of those talks were subsequently revised for publication and have provided the basis for some of what is in this volume, although they have been much revised and expanded. In particular I have drawn upon three essays in the writing of this book: "The Emergence of STS as an Academic Field," *Research in Philosophy and Technology* 9 (1989): 287–301; "The Warp and Woof of Science and Technology Studies in the United States," *Education* 113 (Spring 1993): 381–91, 352; and "Of Frogs, Princesses, and Engineering: A Possible Role for Science, Technology and Society Programs in Education for Sustainable Development," *Sostenible? Tecnologia, desenvolupament, sostenible i desequilibris*, ed. Josep Xercavins i Valls (Barcelona: Universitat Politecnica de Catalunya i Generalitat de Catalunya, Departament de Medi Ambient, 1997): 335–58, 538–53. Other related essays upon which I have also drawn are so noted in the relevant footnotes to each chapter.

I also owe a debt of deep intellectual gratitude to Steve Goldman and Jameson B. Powers for their many years of collegial support in the form of thoughtful conversation, which has led on a number of occasions to coauthorship of essays related to STS educational issues. Both Carl Mitcham and Steve Goldman have read this complete book in manuscript form, as have Paul Durbin and Rudi Volti. I would also like to thank Greg Kunkle for his reading of portions of the book. Wilhelm Fudpucker has constantly prodded me to think, in his words, "outside the box." Without their insights I would have made many more errors of both omission and commission than I am sure are still embedded within what you will read. For this I thank them, while holding them blameless for the problems that remain. I also extend my gratitude to Mary Jo Carlen and Cathy Barrett for their many years of support in coordinating the day-to-day running of the STS Program office at Lehigh, thereby allowing me to sequester the time to read and write. The editorial staff at Rowman & Littlefield and Margaret Trejo and her colleagues at Trejo Production have my deepest appreciation for transforming the original manuscript into this book. Finally, I want to thank the many individuals whose work constitutes the STS field. Without their efforts, insights, and intellectual shoulders upon which to stand, such a work as this would not be.

Notes

1. I am indebted to my Lehigh colleague Steven Goldman for the phrase "ideas, machines, and values" as a way to alternatively characterize the relationships among science, technology, and society.

2. Bruno Latour adopts the term "technoscience" in *Science in Action: How to Follow Scientists and Engineers through Society* (Cambridge: Harvard University Press, 1987), especially 174–75, but also see his more recent comments "From the World of Science to the World of Research," *Science* 280 (10 April 1998): 208–9, where he notes that it is no longer satisfactory to speak of an ideal science somehow separated from society, but rather of a "collective experiment" in which we are all engaged. Tsou Chen-lu, "Science and Scientists in China," *Science* 280 (24 April 1998): 528–29.

1

The Historical Emergence of STS as an Academic Field

If one observes how thoroughly our lives are shaped by interconnected systems of modern technology, how strongly we feel their influence, respect their authority and participate in their workings, one begins to understand that, like it or not, we have become members of a new order in human history.

— Langdon Winner, *The Whale and the Reactor*

Science, Technology, and Society (STS) first emerged as an explicit academic field of teaching and research in the United States in the 1960s. The emergence has a deep historical background in both the modern attempt to transform society through the pursuit of science and technology (the Enlightenment) and the critical reaction to this project (Romanticism). Previous moments in this cultural conflict included the emergence of sociology (the "scientific" study of society) and the history and philosophy of science (society's attempt to comprehend its own creation).

During the mid-1960s, however, views regarding the science-technology-society relationship took on a new form, in large part reflecting a perceived need for a more complete understanding of the societal context of science and technology. STS, especially in the United States, but elsewhere as well, emerged in a period of widespread social upheaval, itself a reaction in part to the social-cultural quiescence of the 1950s. Scholars and more activist critics alike began to raise doubts about the theretofore largely unquestioned beneficence of science and technology that had become the post–World War II consensus. It was as though people had awoken to the fact, as Winner put it some years later, that we were "members of a new order in human history."

Activist groups claiming to speak on behalf of the public interest in such areas as consumerism, civil rights, and the environment, together with protest demonstrations against the Vietnam War, multinational corporations, nuclear power, and so on, set the tone for much of the general context of the period. Within this context there emerged a critique of the idea of progress that by U.S. standards was quite radical. Following a relative collapse in the mid-1960s of a twenty-year-long, direct translation of science and technology into prosperity for the American working class, there emerged the recognition that it was also becoming necessary to cope in practical terms with assessing the value of societal expenditures on science and technology, especially in the face of an accumulated burden of negative impacts. Voices began to question whether science and technology were the unalloyed blessings that society had generally come to believe they were. Both intellectuals and more widely read authors from a variety of perspectives suggested there were negative implications associated with those blessings long assumed to be the primary legacy of science and technology.

Among more popular writers, Rachel Carson, in her 1962 book *Silent Spring*, raised serious questions about the hazards associated with chemical insecticides such as DDT and in many ways helped to crystalize the contemporary environmental movement. At roughly the same time, consumer activist Ralph Nader's 1965 exposé, *Unsafe at Any Speed*, claimed to document the dangers of the Corvair car, and by extension more broadly criticized the cavalier attitude of the automobile industry toward consumers. Like Carson with the environment, Nader played a key role in the galvanizing consumer movement. The 1972 publication of the Club of Rome's *Limits to Growth* and a United Nations–sponsored Conference on the Human Environment held in Stockholm in the same year further reflected public engagement with STS issues and concerns.[1]

The emergence of increasingly sophisticated social movements helped to form the original background of the emergence of STS. Initially these groups included a politically aggressive environmental movement willing to participate in civil disobedience typified by the 1970 Earth Week, during which U.S. Senator Vance Hartke was quoted as saying, "A runaway technology whose only law is profit, has for years poisoned our air, ravaged our soil, stripped our forests bare, and corrupted our water resources"; nuclear power protest organizations including the Abalone and Clamshell Alliances, which respectively fought Pacific Gas and Electric to a standstill over Diablo Canyon for seventeen years and effectively brought construction of Seabrook I to a halt; and the outcry against, and ultimate defeat of, the ABM, the SST, and the use of fluorocarbons in aerosol cans in the early 1970s. Activist groups formed to express concern regarding molecular biology and genetic engineering research, for example, the 1975 Asilomar Conference that resulted in the extraordinary proposal for voluntary restraints on recombinant DNA research, and the 1976 public debates in Cambridge, Massachusetts, over safeguarding research then being conducted at Harvard University. Similarly the 1983 Machinists Union's call for a "New Technology Bill of Rights" that demanded some control over the work process reflected labor's concerns with the impact of new automation technologies on job security, worker safety, and skills reduction.[2]

Similar kinds of STS concerns and responses have of course continued into the present. For example, numerous grassroots environmental groups in the United States and more formalized "green parties" in Europe have come, if not to typify, certainly to play key roles in the environmental arena. In this way, the 1987 World Commission on Environment and Development meeting, chaired by Norway's Gro Brundtland, issued a report, *Our Common Future*, that identified the need for, and focused attention on, "sustainable development" as a way to bridge environmental concerns and developing world concerns regarding economic viability.[3] This was a theme continued and expanded upon at the United Nations–sponsored 1992 Earth Summit in Rio de Janeiro, which yielded among other things an international treaty for the protection of biodiversity. On a much different front, the widespread public exposure, at least within the public news media, to the issues of cloning as represented by the British success with the sheep named Dolly, has led to important discussion and, in the cases of the European Community and the State of California, formal resolutions against human cloning.

Included among the varied political responses to this new public perception regarding science and technology were the establishment by the U.S. Congress of the National Highway Transportation Safety Administration (1966), the Environmental Protection Agency (1969), and the Occupational Safety and Health Administration (1970); the passage of the Clean Air and Clean Water Acts (1970, 1972); and the creation of the now defunct Congressional Office of Technology Assessment (1972). The EPA was created with the related requirement that Environmental Impact Assessments be conducted on all projects involving the federal government or federal monies, and many individual states subsequently enacted similar legislation. OSHA was a response to the impact of technological development in the workplace. The creation of OTA—to say nothing of the emergence of a whole new field of endeavor with its own methodologies, specialist practitioners, and professional societies and journals—was a direct response by Congress to the desire both for technical advice independent of the executive branch and for an effort to anticipate more fully the societal impacts of technology.[4]

In the area of energy, the severing of the regulatory functions from the promotional aspects of the Atomic Energy Commission by the creation in 1975 of the Nuclear Regulatory Commission was likewise a response to increasing concern at all levels regarding the conflict inherent in having both the promotional and regulatory sides of nuclear power contained within the same agency. While some caustic voices might say these were very safe ways for the establishment to respond to valid criticisms without opening up the whole process of scientific and technological decision making to public scrutiny, it is certainly fair to say that the U.S. government has become more sensitive to the societal context of science and technology.

European responses, while not directly parallel to the American experience, nonetheless, reflected similar concerns. In Britain, Derek de Sola Price's 1963 study, *Little Science, Big Science*, prompted debates over what seemed to be a potentially disastrous exponential growth in government funding of science and led to calls for a "science of science." Among the responses was the 1965 formation in London of the Science of Science Foundation. Societies for "Social Responsibility in Science" also appeared in England and elsewhere at about the same time.[5]

Although they were not institutionalized until somewhat later, Denmark also began to pursue technology assessment studies at the level of political culture. This occurred first within the context of labor unions in the late 1970s, then with the Danish Social Science Research Council's establishment of a subcommittee in technology and society in 1982, and then with the creation of a Board of Technology under the auspices of the Danish Parliament three years later.

In contrast to what Lars Fuglsang calls Denmark's bottom up "outside" response to technology, Sweden developed a more top down "corporatist" model, in which debates about technology and "working life" have been frames. Thus, in the mid-1970s the Swedish Parliament widely debated and eventually passed a law on "co-determination in working life" and established a Center for Working Life located in Stockholm in 1976. The goal was to allow Swedish workers to participate more extensively in the planning and organization of the work process, especially as it is affected by scientific and technological change. The Swedes also established a Secretariat for Future Studies, with a charge "to conduct critical projects with technology assessments."[6]

A final example of the range of political responses to the societal implications of science and technology was the establishment in the Netherlands of so-called science shops in which government-supported scientists and engineers provided information and "expert opinion" free of charge to any community group, trade union, or public-interest organization willing to use the information in their work. Collectively these European developments reflected similar concerns regarding science and technology to those that had motivated the U.S. response.[7]

Other more specialized aspects of this increased awareness and concern included the creation of the Ethics and Values in Science and Technology (EVIST) Program (now called the Societal Dimensions of Engineering, Science, and Technology Program—SDEST) within the National Science Foundation (NSF). SDEST, which also includes Science and Technology Studies and the Research on Science and Technology Program (RST), "focuses on improving knowledge of ethical and value dimensions of science, engineering, and technology, and on improving approaches and information for decision making about investment in science, engineering, and technology."[8] The National Endowment for the Humanities (NEH) similarly created the Program in Science, Technology, and Values (now Humanities, Science, and Technology—HST). In a similar way the American Association for the Advancement of Science (AAAS) has identified itself as a partner in trying to reach a better understanding of STS issues through special programs and committees, such as its Directorate for Science and Policy Programs and the Committee on Scientific Freedom and Responsibility (CSFR, established in 1975), annual meeting sessions devoted to STS topics, and the sponsorship of relevant research projects and surveys, such as the Committee on Science, Engineering, and Public Policy's annual *Science and Technology Policy Yearbook*.[9]

Scientists and engineers had also expressed their own questions regarding the course of technoscientific development, partly out of concerns regarding the Vietnam War, when in late 1968 to early 1969 they established the Union of Concerned Scientists (UCS). Perhaps drawing on the tradition of the Federation of American Scientists (FAS, established in 1945), which had emerged out of concerns resulting from the implications of the Manhattan Project, the UCS is committed to "combine rigorous scientific research with public education and citizen advocacy to help build a clean, healthy environment and a safer world." Numbering today some 70,000 members, the UCS is far larger than all the major STS organizations combined. Closely related to the FAS was the 1945 establishment of the *Bulletin of Atomic Scientists,* which continues publication today with a circulation of over 250,000 readers. Its focus is the discussion of science, especially as related to nuclear issues, in an international context. Somewhat more recently, scientists and technical people in the various computer and information technology fields who were becoming increasingly uneasy about the implications of their work joined forces to create Computer Professionals for Social Responsibility (CPSR, established in 1983), an organization devoted to exploring computer-related societal implications in such areas as military concerns, privacy, civil liberties, and the role of information technology in the workplace.[10]

Other professional organizations joined the effort through the establishment of special interest divisions such as the American Society for Engineering Education's Engineering and Public Policy Division and the Institute for Electrical and Electronics Engineers' Society on Social Implications of Technology, which publishes its own *Technology and Society Magazine* and holds a special annual meeting. Sigma Xi, the national scientific honorary society, created a Science, Technology, and Society group charged with focusing precisely on such issues. In 1984 the National Academy of Sciences (NAS), the National Academy of Engineering (NAE), and the National Institutes of Health (NIH) began to jointly publish a new journal entitled *Issues in Science and Technology* devoted to expanding and raising the quality of the national debate on policy issues involving science, technology, and health. The journal continues today under the auspices of the NAS. The almost simultaneous establishment of such agencies, professional societies, and publications indicated the extent to which questions regarding science and technology were taking hold of society and the seriousness with which such issues were being taken at this point in time, both from within the technoscientific community and by the more external academic and public realms.

All such developments reflected an increased interest in the complexities of modern science and technology in contemporary society and attempts to bring to bear a more interdisciplinary approach for understanding not only the obvious benefits of scientific technology but also the often previously ignored negative side effects. In addition to more popular critics, intellectuals from a variety of perspectives extended to the public and the academic world the argument that science and technology were inherently value-laden and often, if not always, problematic in terms of societal impact. Among the most widely read such works were the Frenchman Jacques Ellul's *La Technique l'enjeu du siècle* (1954), which in French means "the gamble or wager of the century," but which, in the 1964 English-language edition, was translated into the somewhat milder *The Technological Society*, and the American Lewis Mumford's two-volume *The Myth of the Machine* (1967 and 1970). Ellul presented a critique of "technique"—"the totality of methods rationally arrived at"—while Mumford assessed what he called the "megamachine," a term denoting the all-encompassing power of modern science-based technology.[11]

Perhaps the most influential intellectual precursor of the STS movement was C. P. Snow. Trained as a scientist, Snow first began to describe the gap between the scientific and literary cultures in a series of novels. Then, in his now-famous 1959 Cambridge University Rede Lecture, he posited the existence of a widening split between "two [noncommunicating] cultures" in society—one composed of scientists, the other of humanists. Snow did recognize that between these two cultures "there are all kinds of tones of feeling on the way," including technology and engineering, and even that of the social sciences, which he suggested was "becoming something like a third culture." Nonetheless, his "two cultures" metaphor did much to shape (and in many ways still serves as a reference point for) discourse within the STS field.[12]

Coterminous with the political and interdisciplinary intellectual responses were cognate changes within a number of traditional disciplinary academic fields as well. Evolving in large part out of the work of such scholars as Thomas Kuhn, John Ziman, and J. D. Bernal, historians, sociologists, and philosophers of both science and technology increasingly moved away from internalist-oriented subdisciplines to progressively more externalist or "contextual" interpretations. This shift, which will be examined in detail in the next chapter, was expressive of the same intellectual and social forces that precipitated the more avowedly interdisciplinary approach of STS. Independent of approach, however, all such developments reflected an increased interest in the complexities of modern science and technology in contemporary society.

Something of a pendulumlike swing of attitudes regarding science and technology has occurred during the course of the development of the STS field. Responding to the largely uncritical stance of the 1950s and 1960s, the tenor of much of the early STS literature was antiestablishment and highly critical in tone, and this was reflected in much of the first generation of STS course work being taught in numerous programs during this period. The initial focus, often coming from engineers and scientists themselves, was frequently directed toward educating science and engineering students about the "true" societal impact of their work. Many of the earliest STS courses and curriculum programs emerged at institutions with engineering colleges and sometimes within those colleges themselves. It was as though STS courses were for adding a cultural veneer to the "coarse" surface of a technical education.

Not unexpectedly, liberal arts students were just as interested in such questions, and very quickly a second generation of STS course work emerged, aimed more generally at all students. This second generation took as its approach a social process interpretation of science and technology. Both were seen as shaped and influenced by societal values, which were, in turn, affected by scientific knowledge and technological values. These developments, taking place in the mid- to late 1970s, corresponded closely with the emergence of a science and technology studies approach to STS and reflected in part an attempt to rise above the fray of a simplistic pro-con debate regarding the merits and demerits of science and technology.

Then during the 1980s, the STS community moved beyond this social content analysis of science and technology, to the design of courses and programs aimed at developing "literacy" on the part of liberal arts students *in* technology, rather than *about* technology. The aim here regarding technology was somewhat parallel to what liberal arts students are expected to learn vis-à-vis science and mathematics. Typical of "literacy" developments during the 1980s were the formation of the Council for the Understanding of Technology in Human Affairs; the emergence of the Alfred P. Sloan Foundation's New Liberal Arts Program, which has produced an extensive series of books, monographs, and extended syllabi; and the holding of a series of annual Technology Literacy Conferences now coordinated under the auspices of the National Association of Science, Technology and Society.[13]

During the latter part of the 1980s and through the mid-1990s, a subsequent interpretative swing toward a "contextualist" or "social constructivist" interpretation has, among many STS scholars at least, led to a view of science and technology in which they are seen as the products not so much of an objective "out thereness," but rather depend upon socially determined causal factors, albeit constrained by material factors in nature. One of the more influential scholars in this regard has been Bruno Latour. In his *Science in Action* (1987), he argues that, to properly understand what he calls "technoscience," one must examine scientists "in action," before the discoveries and inventions become widely accepted or "black boxed." At its most pronounced, this is an extreme view, and one not shared uniformly within the STS community. Nonetheless, it has occasioned a counterresponse, especially among scientists and engineers who want to maintain the objective "reality" of nature and of science and technology in the face of what they see as a misguided "radical" relativism. Perhaps most illustrative of this side of the recent debate was the 1994 publication of a book by Paul Gross and Norman Levitt entitled *Higher Superstition: The Academic Left and Its Quarrels with Science*, in which the authors argued vehemently against the perceived antirealist stance of at least some constructivist STS scholars.[14]

The pendulumlike swing of attitudes regarding science and technology that has occurred during the course of the development of the STS field seems to have damped. In general, the notion of being "pro-" or "anti-" science and technology is not particularly helpful. Very few people today, when expressing criticism of science and technology, mean to suggest doing away with them completely, which would presumably be the logical outcome of an "antitechnology" position. Nonetheless, strong supporters of traditional science and technology are often described as "pro-," while critics are pejoratively criticized as "anti-" science and technology. The latter, especially, makes little sense and would be somewhat akin to calling art critics "anti-art."[15] At the same time, just because we better understand science and technology in their societal context, we cannot afford to lapse into uncritical acceptance. It is precisely this concern that motivates feminist, antiracist, and postcolonial scholars to expose the nonneutrality of science and technology. For they recognize that there is still an element in the science and technology literacy enthusiasm of "if you only understood us better, you would love us more." We must be very careful, in the words of Langdon Winner, to avoid "HSTS—Hooray for Science, Technology and Society."[16]

Certainly it can be said that, on balance, the STS field has moved far beyond any early tendencies toward a simplistic black-and-white image of science and technology in society that it may have had in some quarters to a more sophisticated understanding of the STS relationship. Today STS views science and technology as complex enterprises taking place in specific historical and cultural contexts. What has emerged is a consensus that while science and technology do bring us numerous positive benefits, they also carry with them certain negative impacts, some of which were perhaps unforeseeable, but all of which reflect the values, views, and visions of those in a position to make decisions regarding the scientific and technical expertise within their domain. The central mission of the STS field to date, then, has been to convey a social process interpretation of science and technology. In this view, science and technology are seen as complex enterprises in which cultural, political, and economic values help to shape the technoscientific process, which in turn affects those same values and the society that holds them.

To assist in carrying out that mission, numerous STS programs have come into being during the past three decades. While the specific number is not clearly known, and some have fallen along the wayside, the number of full-fledged programs in the United States numbers nearly one hundred, with perhaps a similar number in Europe. Equally important are the hundreds of individual courses and groups of courses, which, while they cannot be considered programs in the fullest sense, certainly complement the more formally established programs. Similar program and course development has also taken place in Japan, China, Canada, Australia, and several Latin American nations.

What were some of the earliest landmarks, at least in the United States, in this approximately three-decades long development?

The first major effort was the Harvard University Program on Technology and Society funded in 1964 by a five million dollar grant from IBM. Its primary purpose was to "undertake an inquiry in depth into the effects of technological change on the economy, on public policies, and on the character of society, as well as into the reciprocal effects of social progress on the nature, dimension, and directions of scientific and technological developments."[17] Although prematurely disbanded, primarily as a result of bureaucratic infighting, the program did produce a number of studies, books, articles, and bibliographical works culminating in director Emmanuel Mesthene's final report in 1972.[18]

Subsequently, other programs began to emerge with a curriculum orientation. One of the first was the Science, Technology, and Society program at Cornell University, which appeared in 1969 at least in part as a response to campus unrest and the need to develop "interdisciplinary courses at the undergraduate level on topics relevant to the world's problems."[19] That program has since evolved in terms of focusing more extensively on the intellectual study of science and technology, especially at the graduate level, as reflected in its current status as a Department of Science & Technology Studies. Today it is one of the leading STS Ph.D.-granting programs in the United States. Another important early program—the Science, Technology, and Society Program at Pennsylvania State University—emerged out of a "Two Cultures Dialog" begun in 1968–1969. It solidified about 1971, under the influence of the Cornell program. For many years it served as the host institution for the National Association for Science, Technology and Society.

Evolving in a different pattern, but with similar motives, was the 1972 Humanities Perspectives on Technology effort at Lehigh University under a curriculum development grant from the National Endowment for the Humanities. In 1979 it was renamed the Science, Technology and Society Program to bring its title into alignment with the more generic name then coming into vogue across the field. The aim of the original Lehigh program was to "create educational experiences which bring humanistic perspective to the application and evaluation of technology."[20] While Lehigh's program has also grown to include both a modest level of graduate education as well as a somewhat broader contextual focus, it has largely remained true to its original undergraduate, issue-oriented educational mission.

Somewhat later, in 1977, a number of science and technology studies activities at MIT coalesced with the formal establishment of the Program in Science, Technology, and Society. Its aims were "to explore the influence of social, political and cultural forces on science and technology, and to examine the impact of technologies and scientific ideas on people's lives."[21]

These goals, then, as well as those of a host of additional programs too extensive to list individually,[22] reflected a desire to expand and deepen our conceptualization of the workings of science and technology, so as both to understand their societal impacts and to offer insights into better ways of controlling and directing them as societal forces. The rise of undergraduate major programs, for example, Lehigh's STS Program and Wesleyan University's Science in Society Program, and more recently the emergence of graduate degree programs, such as those at Cornell and MIT, as well as a special focus on graduate work in the public policy area, typified by Washington University's Department of Engineering and Policy and a second MIT Technology and Policy Program, to name but two, reflects a sophistication and maturation only initially dreamed of in the late 1960s and early 1970s. The development of such policy-oriented programs further reflects a practical application of STS and reinforces the notion of public involvement in an age in which we are trying to exert stronger and more deliberate social and political control of science and technology.

Taken together these developments suggest the seriousness of purpose with which STS has evolved and an appreciation of the complexities of modern science and technology in contemporary society. At least three different interdisciplinary research and educational approaches to STS can thus be identified: a) Science, Technology, and Public Policy, b) Science and Technology Studies, and c) Science, Technology, and Society programs to distinguish among them. Chapter 4 will take up each of these types of programs in more detail and delineate the current state of affairs within the field, but first it is necessary to turn to a discussion of the cognate changes that occurred within a number of related but more disciplinary-focused fields of study. Suffice it to say for the moment that by roughly the mid-1980s STS had formalized as a sophisticated interdisciplinary field of study with the usual accouterments of academic pursuit—formalized departments and programs, professional societies, and scholarly journals. Today it serves as one of the most exciting interdisciplinary nexuses for the essential study of the relationships among science, technology, and society.

Notes

1. Rachel Carson, *Silent Spring* (Boston: Houghton Mifflin, 1962); Ralph Nader, *Unsafe at Any Speed: The Designed in Dangers of the American Automobile* (New York: Grossman, 1965); Donella Meadows et al., *Limits to Growth: A Report for the Club of Rome's Project on the Predicament of Mankind* (New York: Universe Books, 1972). Carson, Nader, and Meadows's Club of Rome group were by no means the only or even the first critics to question science and technology, for John Kenneth Galbraith in *The Affluent Society* (Boston: Houghton Mifflin, 1958, 2d rev. ed., 1969) and *The New Industrial State* (Boston: Houghton Mifflin, 1967, rev. ed., 1971) had suggested that in the industrial state power had shifted from consumers and the marketplace to a "technostructure" within the corporation that controlled technology for the sake of the growth of the organization. He warned of the instability of an economy keyed to production for its own sake. Preceding Galbraith's work was Vance Packard's *The Hidden Persuaders* (New York: D. McKay, 1957; rev. ed., New York: Washington Square Press, 1980) which painted a picture of the advertising industry as a creator of wants, artificially generating consumer demands while glossing over the absence of real choice. Both authors viewed production as driven by production goals, not consumer needs.

2. Hartke quote in Samuel C. Florman, *The Existential Pleasures of Engineering* (New York: St. Martin's Press, 1976), 13; Mel Horwitch, *Clipped Wings: The American SST Conflict* (Cambridge, Mass.: MIT Press, 1982); James D. Watson and John Tooze, eds., *The DNA Story: A Documentary History of Gene Cloning* (San Francisco: W. H. Freeman, 1982); Sheldon Krimsky, "Regulating Recombinant DNA Research," in *Controversy: Politics of Technical Decisions*, ed. Dorothy Nelkin (Beverly Hills, Calif.: Sage, 1979), 227–53; International Association of Machinists, "New Technology Bill of Rights," *Democracy: A Journal of Political and Radical Change* 3 (Spring 1983): 25–27.

3. World Commission on Environment and Development, *Our Common Future* (New York: Oxford University Press, 1987).

4. See Gregory Kunkle, "Early Warning? The United States Congress and Technology Assessment" (Ph.D. diss., Lehigh University, 1995) for an extended analysis of the debates surrounding the creation of the OTA and for an insightful discussion of public perceptions regarding science and technology more generally during this period; and a related article, "New Challenge or the Past Revisited?: The Office of Technology Assessment in Historical Context," *Technology in Society* 17 (Spring 1995): 175–96. Also useful for the period of the 1970s and 1980s is Bruce Bimber's *The Politics of Expertise in Congress: The Rise and Fall of the Office of Technology Assessment* (Albany: SUNY Press, 1996).

5. Derek de Solla Price, *Little Science, Big Science* (New York: Columbia University Press, 1963). David Edge's overview essay, "Reinventing the Wheel," in the Society for the Social Studies of Science's *Handbook of Science and Technology Studies*, ed. Sheila Jasanoff et al. (Thousand Oaks, Calif.: Sage, 1995), 2–23, contains a good discussion of early British developments in STS.

6. Lars Fuglsang, *Technology and New Institutions: A Comparison of Strategic Choices and Technology Studies in the United States, Denmark and Sweden* (Copenhagen: Academic Press, 1993), see especially chapters 10–11; quotation, p. 155.

7. For an introduction to Dutch science shops, see Richard Sclove, "STS on Other Planets," *EASST Review* 15 (June 1996): 3–7. See also chapter 5 of this volume for a brief discussion.

8. NSF announcement—NSF99-82. Information can also be found on line at: <http://www.nsf.gov:80/sbe/sber/sdest/start.htm>.

9. For a useful discussion of the AAAS's 1975 creation of the CSFR, see Carl Mitcham, "Etica sobre y Dentro de Ciencia y Tecnología," *Theoria* 14 (Fall 1999). The current charter of the CSFR "affirms at the outset that scientific freedom is grounded in basic human rights and implies special responsibilities to extend and disseminate knowledge for the good of humanity."

10. For a general discussion of organizations of scientists and engineers committed to the ethical pursuit of science and technology from "within" their own fields and professional organizations, as opposed to concerns such as those expressed by STS from "outside," see Mitcham, "Etica sobre y Dentro de Ciencia y Tecnología" and Mitcham, "Engineers and Scientists as Ethical Leaders in the Technoscientific World," unpublished paper read at the Internacional Menedez Pelayo Tenerife workshop on "Ciencia, tecnología y valores: Reflexiones en visperas del nuevo milenio," San Sebastian, Spain, April 5–9, 1999. Quotations are drawn from these two works. The CPSR was officially formed in California in

1982, and then became a national organization the following year. Further information about CPSR, its historical development , and its many activities and publications, including the *CPSR Newsletter*, can be found at its web site: <www.cpsr.org>.

11. Jacques Ellul, *The Technological Society*, trans. by John Wilkinson (New York: Knopf, 1964); Lewis Mumford, *The Myth of the Machine*, 2 vols. (New York: Harcourt Brace Jovanovich, 1967–1970). Among other quasi-intellectual popularizers who had an important influence on public perceptions regarding the societal implications of science and technology during this same time period were Theodore Roszak, *The Making of a Counter Culture: Reflections on the Technocratic Society and Its Youthful Opposition* (Garden City, N.Y.: Doubleday, 1969) and *Where the Wasteland Ends: Politics and Transcendence in Postindustrial Society* (Garden City, N.Y.: Doubleday, 1972); and Alvin Toffler, *Future Shock* (New York: Random House, 1970).

12. C. P. Snow, *The Two Cultures and the Scientific Revolution* (Cambridge: Cambridge University Press, 1959) and rev. ed. *The Two Cultures: And a Second Look* (Cambridge: Cambridge University Press, 1964), quotations 11, 69–71. I am indebted to the work of Howard Segal for this insight into Snow; see his essay "High Tech and the Burden of History: Or, the Many Ironies of Contemporary Technological Optimism," in *Future Imperfect: The Mixed Blessings of Technology in America* (Amherst: University of Massachusetts Press, 1994), especially 195, 197. Indicative of the ongoing interest in the issues raised by Snow's metaphorical image are Jonathan Cole's recent essay, "The Two Cultures Revisited," in the National Academy of Engineering's journal *The Bridge* 26 (Fall/Winter 1996): 16–21, in which he argues that "the gulf in understanding between scientists and nonscientists may be traceable to an educational system that neglects the historical importance of scientific and technological developments."

13. The relationship of "technology literacy" to STS and liberal arts education is elaborated upon in Steven L. Goldman and Stephen H. Cutcliffe, "STS, Technology Literacy, and the Arts Curriculum," *Bulletin of Science, Technology & Society* 2, no. 4 (1982): 291–307; Cutcliffe and Goldman, "Science, Technology, and the Liberal Arts," *Science, Technology, & Human Values* 10, no. 1 (Winter 1985): 80–87; Cutcliffe, "Understanding Science, Technology, and Engineering: An Essential Element of Cultural Literacy," *Federation Review: The Journal of the State Humanities Councils* 8, no. 4 (July/August 1985): 10–15; and Goldman, "The Warp and the Woof," *The Weaver* (1985), p. 2. See also Barrett Hazeltine, "Past Efforts in Technological Literacy—CUTHA," in *Technology Literacy Workshop Proceedings*, ed. Russel C. Jones, Accreditation Board for Engineering and Technology, the Association of American Colleges, National Science Foundation (Newark: University of Delaware, 1991); James D. Koerner, ed., *The New Liberal Arts* (New York: Alfred P. Sloan Foundation, 1981); and Samuel Goldberg, *The New Liberal Arts Program: A 1990 Report* (New York: Alfred P. Sloan Foundation, 1990).

14. Bruno Latour, *Science in Action: How to Follow Scientists and Engineers through Society* (Cambridge: Harvard University Press, 1987); Paul Gross and Norman Levitt, *Higher Superstition: The Academic Left and Its Quarrels with Science* (Baltimore: Johns Hopkins University Press, 1994).

15. David Dickson, *The New Politics of Science* (Chicago: University of Chicago Press, 1980), 6; Langdon Winner, *The Whale and the Reactor: A Search*

for Limits in an Age of High Technology (Chicago: University of Chicago Press), 1986), xi. Bruno Latour similarly argues that just because people study a subject matter doesn't mean they are automatically against it. He asks, "are biologists anti-life, astronomers anti-stars, immunologists anti-antibodies?" *Pandora's Hope: Essays on the Reality of Science Studies* (Cambridge: Harvard University Press, 1999), 2.

16. Langdon Winner, "Conflicting Interests in Science and Technology Studies: Some Personal Reflections," *Technology in Society* 11 (1989): 436.

17. Quoted in Albert H. Teich, ed., *Technology and Man's Future*, 4th ed. (New York: St. Martin's Press, 1986), 3.

18. Emmanuel Mesthene, *Harvard University Program on Technology and Society, 1964–1972: A Final Review* (Cambridge: Harvard University, 1972).

19. Franklin A. Long, *First General Report, Cornell University Program on Science, Technology, and Society* (Ithaca, N.Y.: Cornell University, 1971), 2.

20. Edward J. Gallagher, *Humanities Perspectives on Technology, Annual Report Year Five, 1976–1977* (Bethlehem, Pa.: Lehigh University, 1977), iii.

21. *Program in Science, Technology, and Society* (Cambridge, Mass.: MIT, 198), 3.

22. Unfortunately no current comprehensive guide to STS programs exists. However, several surveys of STS programs conducted in the mid-1970s are helpful for understanding the formative period: Ezra D. Heitowit, Janet Epstein, and Gerald Steinberg, *Science, Technology and Society: A Guide to the Field* (Ithaca, N.Y.: Cornell University Program on Science, Technology and Society, 1977), and *EVIST Resource Directory: A Directory of Programs and Courses in the Field of Ethical Values in Science and Technology*, Document 78-6 (Washington, D.C.: American Association for the Advancement of Science, 1978). A brief follow-up to the Heitowit surveys was conducted by Rustrum Roy and Joshua Lerner in 1982–1983: Roy and Lerner, "The Status of STS Activities at U.S. Universities," *Bulletin of Science, Technology and Society* 3, no. 5 (1983): 417–32, and, much more recently, the Directorate for Science and Policy Programs of the American Association for the Advancement of Science has published the third edition of Albert H. Teich, ed., *Guide to Graduate Education in Science, Engineering and Public Policy* (Washington, D.C.: AAAS, 1995), which includes material on twenty-eight U.S. graduate degree-granting programs and another fourteen programs outside the United States in this particular facet of STS. The National Association of Science, Technology and Society has issued under the editorship of Carl Mitcham and Stephen H. Cutcliffe a second edition of its *STS Directory* (University Park, Pa.: NASTS, 1996), an admittedly incomplete survey of some sixty U.S. and international STS programs. Most recently, the European Commission has made available Paul Wouters, Jan Annerstedt, and Loet Leydesdorff, *European Guide to Science, Technology, and Innovation Studies* (Brussels: European Commission, 1999), which is available electronically at: <http://www.cordis.lu/tser/home.html>. Although useful entrees to many of the major STS programs in the United States, Britain, Europe, and some other parts of the world, these surveys do not track the many hundreds of individual courses or clusters of courses that also contribute to the academic vitality of the STS field.

2

Societal Contextualization in the Philosophy, Sociology, and History of Science and Technology

Technology is neither good, nor bad; nor is it neutral.

— Melvin Kranzberg, "Kranzberg's Laws"

[T]echnics . . . does not form an independent system, like the universe: it exists as an element in human culture and it promises well or ill as the social groups that exploit it promise well or ill.

—Lewis Mumford, *Technics and Civilization*

The emergence of Science, Technology, and Society as an academic field of study some three decades ago reflected, in large part, a perceived need for a more complete understanding of the societal context of science and technology. As noted in chapter 1, one outcome of the tensions of the 1960s and early 1970s was a critique of science and technology as analysts began to focus on the negative externalities that they perceived to be affecting the modern world. Out of this turmoil emerged an interdisciplinary and issue-oriented, activist field of study that sought both to understand and to respond to the complexities of modern science and technology in contemporary society.

At approximately the same time that STS was emerging, cognate changes in the approaches of a number of more narrowly discipline-oriented academic fields also took place. Illustrating and contributing to these changes was the enormous influence of Thomas Kuhn's *The Structure of Scientific Revolutions*, first published in 1962.[1] More or less independently of each other, philosophers, sociologists, and historians of science and technology moved away from internalist studies to more contextual interpretations. The common denominator across these six fields was a critique of traditional notions of "objectivity" within scientific and technological knowledge and action, a critique that emphasized the value-laden contingent nature of these activities. This is why Melvin Kranzberg argued that technology was never "neutral."[2] For most scholars this meant not an outright denial of the "reality" of nature or of artifacts, but only an insistence that our understanding of nature and of science and the creation of technology are societally mediated processes. This chapter will explore briefly some of the key developments in each of six distinct fields.

Philosophy of Science

The explanatory power of the natural sciences, and especially physics, became increasingly powerful during the early twentieth century, with many viewing them as an important motive force in the creation of modern society. Philosophers as well as others readily took note and correspondingly moved away from their earlier broad metaphysical concerns to focus on more concrete problem areas. It was thus that philosophy of science became one of the earliest "philosophies of . . . " with the Philosophy of Science Association being officially founded in 1934. For most of the middle third of the twentieth century, philosophers of science espoused a positivist model of science. In this view, scientific facts and theories are seen to represent, usually in logical or mathematical terms, something about the real external world, which, although mediated by human experience, is not fundamentally shaped by societal factors. Thus, positivism is concerned primarily with the logical structure of scientific explanation, beginning with perceptual experience, not the world per se. For this reason positivists demonstrated little interest in the messy process of scientific inquiry itself, which was deemed largely irrelevant to the discovery of laws that existed irrespective of the observer.[3]

Some positivists such as Carl Hempel and Willard V. O. Quine, however, soon came to question the ability of science to separate out theoretical and observational elements or terms in a scientific theory; rather, they saw terms as being holistically linked and argued they must be accepted as part of an interdependent network. Conversely, it became clear that claims about observable phenomena also depend upon other beliefs and theories held by scientists. This led to N. R. Hanson's argument that all scientific claims are "theory-laden" and his related observation that "People, not their eyes, see."[4] Such a holistic view suggests the theory-laden context of scientific phenomena and terms and hence opens the door to questioning the supposed "objectivity" of science. Quine went further, in what has become known as the Quine-Duhem Thesis, to argue that it is, in fact, impossible to design experiments that, with absolute certainty, will prove or disprove a given theory. Even disconfirming observational evidence can be accommodated by the scientist. In this view, theory is "underdetermined" by the observational evidence. That is, by itself, the evidence cannot determine absolutely the truth or validity of one given theory over some alternative explanation that fits equally well the evidence. It is thus we, not nature, who determine the legitimacy of a scientific theory.[5] It was through this partially opened door to constructivist interpretation that Thomas Kuhn now stepped with the publication of his work on scientific revolutions.

Although by no means the first or only scholar to suggest the contextual nature of science, Kuhn dramatically reoriented the way philosophy of science was focused with the publication of *The Structure of Scientific Revolutions*. A shift occurred—away from the abstract analysis of well-established theories and toward a historical interpretation of the actual process of doing science. Kuhn and others, upon looking closely at the historical evidence, discovered that scientific theory formation was not as rational nor as progressively accumulative as the positivists imagined it to be; in fact, scientific practice was messier and far more arbitrary. Although Kuhn himself hardly intended it, his work introduced a relativist interpretation into the philosophy (and the history and sociology) of science, which, in the hands of others and in its most extreme presentation, became an antirealist stance.

Kuhn's analysis suggested that scientific knowledge evolved far more discontinuously, much as does art in the history of that field, than suggested by the positivist model, which depicted an incrementally progressive accumulation of knowledge that increasingly mirrored "reality" or the truth regarding nature. Rather than focusing on the analysis of mature theories and scientific explanation, as had the positivists, Kuhn directed his attention to the everyday practice of science. He argued that science is organized around what he termed "paradigms," or organizing patterns of belief and practice. Although any new field of science goes through a formative period in which there may be theoretical confusion and a number of competing schools of thought, eventually a paradigm that defines the practice of "normal science" within the given domain will emerge. Once that paradigm emerges, the practice of so-called normal science operates within accepted norms and principles and solidifies the new theories by slowly working on solutions to details and minor questions left unresolved by the larger, and necessarily incomplete, paradigmatic framework. Kuhn called such research work "mopping-up operations."[6] In normal science little attention is focused on the unexpected or novel; rather, attention is directed toward puzzles that are assumed solvable within the framework of the paradigm. In this way paradigms also constrain the practice of normal science both by identifying such puzzles and by defining "admissible solutions."

Paradigms are inherently imperfect and necessarily incomplete, however, with the result being that "anomalies"—phenomena not explainable by the theory—will inevitably occur. In the course of conducting normal scientific research, most such anomalies fail to challenge the paradigm and hence, in effect, are ignored. Only when sufficient anomalies occur outside the explanatory framework of the paradigm does a period of "crisis" emerge that challenges the reigning paradigm. When the challenge is too great to withstand, a "scientific revolution" takes place, a period characterized by struggle and controversy, and eventually the replacement of the old with a new paradigm, one that appears to explain the recent anomalies. Another period of normal science can now begin. Science has advanced, but it has done so discontinuously, rather than incrementally as under the positivist view.

Kuhn argues that competing paradigms are, by definition, incompatible with each other and hence the newly emerged normal tradition will be "incommensurable with that which has gone before."[7] How a scientist chooses (or does not choose) to go along with the new paradigm depends not on the basis of some rationally neutral judgment, which cannot exist, but on the basis of what Kuhn variously refers to as a Gestalt switch regarding some sense of what constitutes "better" explanations, "more interesting" problems, or "novel" predictions. Such a change in world view is influenced by "societal" factors—education, research funding control, textbook and journal editorship, aesthetics. Kuhn also argues that new paradigms do not necessarily come closer to representing the "truth" about nature. Science is "a process of evolution *from* primitive beginnings—a process whose successive stages are characterized by an increasingly detailed and refined understanding of nature" but not necessarily "a process of evolution *toward* anything."[8] Again, one might compare this to developments in art, where historical shifts from one school to another represent changes in taste and focus but not necessarily "improvements."

Philosophical relativists and antirealists have seized on this line of argument, positing that what passes for reality is a function of human theorizing about the structure of nature. That a given scientific theory appears empirically successful is either a matter of chance or due to the fact that success has been defined in such a way as to make it adequate, even if not literally true. Although he himself was uncomfortable with these relativist implications, Kuhn's paradigm-oriented interpretation of science cast into doubt the "reality" of nature, thereby opening the door further to antirealist views, which have generally been characterized as "social constructivism."[9]

Sociology of Science

While a constructivist perspective and approach has come to influence most of the disciplines involved in science studies, its greatest influence has been in the area of sociology. Prior to about World War II, sociology of science was closely linked to positivist philosophical views of knowledge that viewed science as lying outside societal influences, and hence not appropriate for sociological study. Perhaps foremost among those few sociologists who began to look more closely at science during the 1940s and 1950s was the functionalist, Robert Merton. While accepting science as a method of objective inquiry, and hence excluding scientific knowledge itself from sociological inquiry, Merton and like-minded colleagues did seek to explain the functioning of science institutionally and organizationally. Merton viewed science as operating within a set of norms—universalism, communality, organized skepticism, disinterestedness—a collective ethos by which scientists develop an objective understanding of nature and its operations, and are recognized for having done so.[10]

The appearance of Kuhn's work on scientific paradigms in the early 1960s, which in some ways was not that different from Merton's notion of norms in terms of the need for consensus if science is to "advance," nonetheless raised questions of why scientists choose to act the way they do. During the 1970s sociologists increasingly found the notions of "scientific community," and even "paradigm," lacking in full explanatory power, arguing instead that nonscientific "social" causes lay at the heart of the matter. Among the first more empirically minded sociologists to move beyond the Mertonian interpretation were Michael Mulkay, David Bloor, and Harry Collins. Mulkay took the view that scientific knowledge is dependent upon societal context, wherein the interests of scientists, their position within the hierarchy of the scientific establishment, financial considerations, government or corporate support, and so on, will influence that which is deemed "useful." Thus, while he accepts the fact that physical reality "constrains" scientific conclusions, Mulkay argues that it is not the sole determinate of what passes for acceptable science.[11] Going a step further down the relativist path, albeit without denying an independent material world, is the work of the sociologist of scientific knowledge, David Bloor, who advocates what he calls the "strong programme." This approach, sometimes referred to as the Edinburgh School in recognition of the institutional affiliation of Bloor and his colleagues, argues that all knowledge claims—both "true" and "false" beliefs in science—are to be explained by the same social, nonrational reasons. In this view, the content of scientific explanation, not just the conduct of scientists, is constructed. This is why that which is discarded must be treated in exactly the same way as knowledge that is acceptable. To state this argument even more dramatically, for the strong programme, scientific "reality" is not acceptable as a belief explanation over "erroneous," irrational ideas that are subsequently discarded. Both types of belief must be treated equally in sociological terms; neither is necessarily better. To accept the latter would be a "weak" explanation for the development of scientific knowledge. Thus, for Bloor and like-minded scholars, scientific knowledge is socially influenced; this relativistic approach has also become known as the "interests" approach.[12]

Sociologists have frequently found that focusing on scientific controversies is particularly revealing of the "interests," both social and political, held by scientists that influence the outcome of their work. By way of example, Harry Collins has examined debates over topics as diverse as the building of TEA lasers, gravitational waves, and parapsychology to show that scientists, far from utilizing the rigorous "scientific model" of Popperian experimental "falsification," in fact utilize a wide range of negotiated and socially mediated beliefs to reach consensus in a given controversy. The work of Trevor Pinch on solar-neutrino detection takes a similarly relativistic approach.[13] Some sociologists of science have recognized, by their own criteria, that their studies of scientific knowledge are similarly constructed or "interpretative." Hence they are no more privileged or "real" than other explanations. Similarly constructed then, their conclusions about scientific knowledge must be equally open to deconstruction through what they often term "self-reflexive" analysis. Such an approach can contribute to an increasingly relativistic worldview, and by extension it has contributed to a social criticism of science and its negative implications more broadly.[14]

Drawing on this evolving sense that even their own studies were sociologically mediated, and taking a cue from field anthropologists, a number of sociologists of science determined that a deeper, more revealing understanding of science would come through studying scientists in action. They believed it far more revealing to analyze what scientists do and say while they are doing it, rather than refraining until, influenced, however subtly or unknowingly, by various interests and societal influences, they have reified their conclusions regarding acceptable knowledge claims. To that end, in the mid-1970s, they began to pursue scientists into their laboratories both through "ethnographic" study and through "discourse analysis," the study of what scientists actually say and publish. Sometimes referred to collectively as EDA, such ethnographic and discourse analysis must be evaluated reflexively. While not all sociologists have found such a reflexive approach fruitful, in part because it tends to divert attention away from the object of study, from the scientific activity itself, it has, nonetheless, offered a number of useful, even if controversial, insights. Probably the most important, widely cited, and, for some at least, problematic of such studies has been the work of the French philosopher and sociologist Bruno Latour.

In the mid-to-late 1970s, Latour spent an extended period of time in ethnographic observation of scientists at the Salk Institute for Biological Studies in La Jolla, California. His findings were published in a volume coauthored with Steve Woolgar entitled *Laboratory Life: The Construction of Scientific Facts*. In what was, in effect, the first such detailed case study, the authors' approach was to observe and record ethnographically what the scientists did and said, rather than to take at face value what the scientists said *about* what they did. What Latour and Woolgar conclude, and what Latour elaborates upon more fully in a later, more generalized, study, *Science in Action*, is that all scientific facts are "socially constructed." That is, what counts as scientific fact, as nature, is not uniquely and logically the result of some objective reality "out there." It is instead the result of a collective process of persuasion, equipment and laboratory building, paper publishing and citations—in short, network building. All of this eventually comes together, allowing scientists to reach "closure" and to "black box" what the "facts" really are. Such agreement regarding nature, scientific theories, and even technology occurs, not because there are facts that are independently "true," but because enough people have become convinced, or in Latour's words "enrolled," regarding their truth. In other words, people come to accept that what scientists say works, in fact, actually does so, and, therefore, they accept it as real. Thus, it is the content of science that defines reality, not the other way around. Latour argues that to gain sociological insight into this process, "We study science *in action* and not ready-made science or technology; to do so, we arrive before the facts and machines are black boxed or we follow the controversies that reopen them." Although this brief summary hardly does justice to the very real influence and contribution of Latour's work, it does reveal the increasingly relativistic line of argument within the sociology of scientific knowledge (SSK). It is a perspective that has aroused a great deal of controversy in some quarters, especially among many scientific and technical people, but also among philosophers and historians, who wish to hold on to a more realist conception of science and technology, even while recognizing their societal embeddedness.[15]

The interest in sociology of science would not really coalesce institutionally until the mid-1970s. Prior to that time, with but a few exceptions such as Merton, Bernard Barber, and Joseph Ben-David, general sociologists expressed little concern for either science or technology as topics of much import. By the late 1960s and early 1970s, however, sufficient interest had emerged that, in the face of a still unenthused American Sociological Association, which required 200 members to establish a special interest section, a group of scholars founded a new and separate Society for the Social Studies of Science (4S) in 1975. Merton served as the first president, reflecting the sociological influence that underlay the organization's founding, but science policy interests have also played a significant role within the purview of 4S, as have history and to a lesser extent other fields as well. Of late, 4S interests have broadened to include technological issues as well as science.[16]

History of Science

Historians of science, and of technology, have also played an important role in the emergence of science and technology studies. First, their analyses of science and technology as worthy subjects in and of themselves have been valuable. Second, they have provided historical grist for numerous sociological case studies and for philosophical reflection.

History of science got its start, at least in the English-speaking world, through the work of William Whewell, master of Trinity College, Cambridge University. Whewell, who wrote on many subjects, focused much of his attention on surveying the history of the exact sciences, especially in his 1837 three-volume study, *History of the Inductive Sciences*.[17] Indeed, it was Whewell who coined the term "scientist" in the course of the founding of the British Association for the Advancement for Science some six years earlier.

While certainly there were many other British and European historians of science working in the nineteenth century, George Sarton, a Belgian scientist and mathematician turned historian, is usually credited with founding the history of science as a modern academic discipline.[18] Before leaving Belgium in 1915 for the United States, and subsequently a position at Harvard University, Sarton had earlier founded and in 1913 published the first issue of *Isis*, "a review dedicated to the history of science," which today continues as the field's leading scholarly journal. It is the official publication of the History of Science Society subsequently founded in 1924 by Sarton and others who saw the history of science as contributing to what he called a "New Humanism." Sarton was much influenced by the French positivist August Comte and his secular faith in the progressive nature of science. As editor of *Isis*, a prolific bibliographer, and author of numerous guides to the field, Sarton was able to imbue countless colleagues and students with a progressive, cumulative view of a growing body of scientific knowledge. Sarton and many of his colleagues tended to focus on key historical moments and "heroic" individuals whom they saw as contributing to and defining science's progressive development. Unlike later successors, this first generation of scholars paid little attention to sociopolitical context.

Despite an interwar years tangent that saw the adoption of certain Marxist influences and insights, such as reflected in the work of Cambridge scientist/historian J. D. Bernal or, earlier, that of German sociologist/historian Max Weber, most historians in the immediate post–World War II era continued to view science as an intellectually abstract and theoretical search for truth.[19] Among the proponents of this perspective was Alexandre Koyré.[20] Inspired by science's obvious and increasing power, historians of science believed the discipline could at once attract students into careers in science, while at the same time promoting the value of the sciences among the broader public. Often drawing on individuals originally trained in science, the history of science, much like the philosophy of science, tended to focus on "internalist" topics centered on well-accepted, major theories, especially in the physical sciences. For example, I. B. Cohen wrote about the emergence of a new physics during the scientific revolution and Thomas Kuhn about Copernican astronomy, while Henry Guerlac focused on the work of Lavoisier and Charles Gillispie on that of Carnot.[21] The now formal discipline of history of science expanded rapidly during the 1950s and 1960s, but by the end of this period some scholars began to question the progressive and linear growth model of scientific development.

Perhaps in part fueled by this very growth and looking for new intellectual outlets, but certainly influenced by larger societal and political issues regarding science and technology—environmental concerns, population and food pressures, nuclear weapons, the Vietnam War—many American and European historians began to look much more closely at the values embedded in science and the objectives underlying its pursuit. To this focal shift, the importance of Kuhn's *The Structure of Scientific Revolutions* cannot be underestimated, any more than it can be for the sociology of science. Following Kuhn's lead, during the decade of the 1970s, a spate of work began to focus on the social roots and context of science. Representative of this burgeoning approach were such books as: Jerome R. Ravetz's *Scientific Knowledge and Its Social Problems* (1971); Gerald Holton's *Thematic Origins of Scientific Thought* (1973); Mary B. Hesse's *The Structure of Scientific Inference* (1974); Charles E. Rosenberg's *No Other Gods: Science and American Social Thought* (1976); and Daniel J. Kevles's *The Physicists* (1978).[22] The mathematician-philosopher Ravetz talked about the ways industrial concerns affected scientific research, in effect, "industrializing" scientific knowledge. Rosenberg discussed the social uses of scientific ideas. Kevles stressed the political dimensions underlying the institutional development of the American physics community.

As the history of science became more social, and sociology of science became increasingly historical, in the latter 1970s and 1980s a number of studies began to draw on the insights and methodologies of both fields. These works sometimes blurred the disciplinary boundaries, a not altogether bad thing from the perspective of science and technology studies. David Edge and Michael Mulkay's study of the emergence of radio astronomy in Britain and the more recent work by sociologist Steven Shapin and historian Simon Schaffer, *Leviathan and the Air-Pump*, represent two examples of this interdisciplinary trend.[23]

Shapin and Schaffer sought to demonstrate the fuzzy nature of scientific conceptualization through a historical examination of the dispute between Robert Boyle and Thomas Hobbes over the vacuum air-pump. At stake was Boyle's support of inductive inference supported by empirical experimentation in science against Hobbes's support of deductive a priori knowledge systems; each distrusted the other's motives and approach to understanding knowledge. Traditionally in the history of science, Boyle's view has been seen as the more scientific and "correct," while Hobbes has largely been dismissed as wrongheaded. Shapin and Schaffer turned this interpretation on its head, by suggesting that Boyle's "victory" in this dispute resulted from a partisan network of support by his colleagues and later proponents on his behalf, his "enrollees" to borrow a term from Latour. The authors argue that with a slight contextual rearrangement of support, Hobbes might well have been victorious. In their conclusion, "The Polity of Science," they emphasize that "it is ourselves and not reality that is responsible for what we know. Knowledge, as much as the state, is the product of human actions." By no means have scholars uniformly accepted the conclusions of Shapin and Schaffer; nonetheless, this widely cited study has been extremely important for the way it has refocused the kinds of questions historians of science ask.[24]

History of Technology

While today it has become fairly commonplace for historians of science to work in a contextual or constructivist mode, this was not always the case. Nor was it always typical for historians of science to express much interest in the history of technology as distinct from science. Indeed, professional historians, as well as the public, often misconceived technology and engineering as being mere applications of more theoretical knowledge. A part of this lack of interest may also have stemmed from a certain disdain for engineering and technology. Nonetheless, despite these misconceptions, there had been a long, even if "minority," tradition within the history profession that expressed interest in technological topics. Many of these early historians interested in technology adopted a progressive stance, idealizing its contributions to societal well-being.

Such idealization often led to a cataloging of contributions from an "internalist" perspective. Thus, as early as 1862 the British biographer Samuel Smiles would eulogize the technical accomplishments of heroic inventors, industrialists, and engineers in his *Lives of the Engineers*, which appeared in numerous editions.[25] In this century just after World War I, a number of British scholars founded the Newcomen Society for the Study of the History of Engineering and Technology and in 1920 began publishing a journal of *Transactions*. Similar in progressive spirit to the work of Sarton in history of science, these early historians of technology, for the most part, adopted an "internalist" or nonsociological mode of presentation. Perhaps typical was the multivolume encyclopedic *History of Technology*. The chief editor, Charles Singer, himself strongly influenced by Sarton, organized the material by industrial process and artifact type with little regard for its sociopolitical context.[26]

There were, of course, exceptions to this deterministic progressive view of technology, perhaps chief among them being the early work of Lewis Mumford and that of Siegfried Giedion. Both Mumford, a regional planner and architectural historian, and Giedion, a historian of art and architecture, viewed technology as a subject worthy of study in itself. Taking a more "holistic" view, they sought to counter the misconception that technology evolves under its own imperatives, somehow separate from human direction and cultural context. It was just such a view that brought Mumford in *Technics and Civilization*, with a somewhat more optimistic outlook than that which would come to characterize his later work, to describe technics not as an "independent system" but as "an element in human culture" that "promises well or ill as the social groups that exploit it promise well or ill."[27]

Despite the contextual insights of writers like Mumford and Giedion, it would not be until the late 1950s that historians of technology would begin to challenge the professional hegemony of the historians of science. Inspired in part by the American Society of Engineering Education, which sought to encourage the history of technology among young engineering students, while at the same time frustrated by the seeming lack of interest among historians of science for such practical considerations, a small group of scholars under the leadership of Melvin Kranzberg, then a professor at the Case Institute of Technology, in 1958 established the Society for the History of Technology (SHOT). The following year the society began publishing *Technology and Culture*, under what would become Kranzberg's long-term editorship. Here the selection of a title is instructive. The title might just as well have been a much more mundane Journal for the History of Technology, except for the fact that right from the beginning SHOT's founders intended a more cultural, or "contextual" approach. Thus, SHOT and *Technology and Culture* have long been "concerned not only with the history of technological devices and processes but also with the relations of technology to science, politics, social change, the arts and the humanities, and economics." Early issues of the journal did tend to include much that might be viewed as "internalistic"; for example, an entire issue in 1960 was devoted to reviews of Singer et al.'s *History of Technology*, and similar national works from several other countries. Very shortly, however, the field had clearly adopted a "contextual" outlook and approach.[28]

An early work dating from this period and representative of the contextual shift was Lynn White jr.'s *Medieval Technology and Social Change* (1962), which explicated the relationship between the technical (e.g., the stirrup) and its societal context (e.g., shock combat and the rise of feudalism and chivalry). Typical of more recent contextual work is that of Ruth Schwartz Cowan. First in a 1976 article in *Technology and Culture* and subsequently in a book-length study, *More Work for Mother* (1983), Cowan explains contextually the seeming paradox between the widespread dissemination of "labor saving" household technologies—irons, vacuum cleaners, washing machines—and the fact that they did not reduce the actual hours of labor expended in household work. The answer lay in cultural and societal factors—reduced numbers of servants, more loads of laundry now that some of the drudgery had been eliminated, and the substitution of tasks such as shopping for goods once delivered directly to the door. For Cowan, the focus of her study is on "social, rather than technical dynamics." Taking a somewhat more technical focus, Thomas P. Hughes, in *Networks of Power* (1983), examined comparatively the technical development of electrical "systems" in the United States, Britain, and Germany between 1880 and 1930, which he found tied as much to political and economic factors and constraints as to anything technical. Although Hughes's work is situated somewhat more closely to the technical end of the contextual spectrum, it nonetheless reflects and addresses the societal context within which his technical systems are embedded. Hughes's work, through his ideas regarding technological systems, has also provided a strong link to the evolving field of the sociology of technology and its focus on networks of both technical and human factors.[29]

Sociology of Technology

The sociology of technology, as distinguished from its scientific cousin, has a much shorter history. Although Karl Marx certainly wrote about technology's role in society, especially with regard to economic systems and the workplace, he was not fundamentally concerned with technology. Such concern awaited the social theorist William Ogburn, whose 1922 *Social Change with Respect to Culture and Original Nature* sought to understand and measure cultural change. He was particularly interested in the process of invention, which he viewed in evolutionary terms, and its impacts on society, to which society subsequently adapted. When technological innovations outstripped society's ability to readily adapt, Ogburn viewed this as evidence of what he called "cultural lag." While Ogburn's deterministic view of technology's impact upon society has been largely rejected, his sociological analysis represents an important early step in social theorizing about technology. Ogburn's work was extended by his colleague and sometime coauthor, S. Colum Gilfillan. Gilfillan's 1935 *Sociology of Invention* focused on invention as a slow incremental process of change. This evolutionary perspective led Gilfillan, together with Ogburn, to become interested in technology assessment and forecasting issues. Together they coauthored several assessment studies, including *The Social Effects of Aviation*. While both men would go on to publish further studies regarding technology, their influence on the subsequent course of academic sociological study dissipated, perhaps due to the rising influence of the Mertonian school of thought and Kuhn's later work on scientific revolutions. Whatever the reasons, the sociology of technology went into something of an intellectual hiatus for a generation, only to reemerge in the 1980s in what some scholars have referred to as "the turn to technology."[30]

To be sure there were in the postwar era numerous studies of technology that might be deemed sociologically oriented; one need only think of some of the work of the French industrial sociologist Georges Friedmann, the French social theorist Jacques Ellul, and the historian and social critic Ivan Illich.[31] Nonetheless, it was not until the appearance of Donald MacKenzie and Judy Wajcman's 1985 edited volume, *The Social Shaping of Technology*, and the even more influential *Social Construction of Technological Systems*, edited by Wiebe Bijker, Thomas Hughes, and Trevor Pinch, two years later, that sociologists seriously began to pay renewed attention to technology. Such a "turn" to technology should not have been surprising, except perhaps to academic sociologists caught up in their enthusiasm for sociology of scientific knowledge (SSK); certainly the historians and philosophers of technology, as well as more activist-oriented STS writers, had been considering the technological as distinct from, yet equally important to, the scientific for some time. Nonetheless, sociologists taking their cue from SSK now sought to understand the ways in which society shaped and "constructed" technology. MacKenzie and Wajcman sought answers to the questions "What has shaped the technology that is having 'effects'?" and "What . . . is causing the technological change whose 'impact' we are experiencing?" Similarly, Pinch and Bijker wanted "to open the so-called black box in which the workings of technology are housed," a theme paralleled in the work of Latour as noted earlier in this chapter.[32]

Not surprisingly, as is the case with SSK, there is a range of approaches to the sociological study of technology, of which three should be briefly mentioned. The first, "systems theory," evolves out of the work of Thomas Hughes and is at least as historical as it is sociological in nature. Drawing on his study of electrical power systems, Hughes argues that technology, at least large-scale systems, should be viewed in terms of "networks" consisting not only of technical artifacts but also of the environment in which they reside—people, institutions, and artifacts function in a "seamless" web. As such networks involve ever larger numbers of people and solve problems—both technical and political—they take on a certain "momentum" of their own, making them increasingly difficult to alter in terms of their societal embeddedness. Although by no means autonomous in Hughes's mind—since differing cultural contexts directly influence the construction of the system, leading, for example, to varied "national styles"—momentum does result in certain impacts on society, exerting what he calls a "soft determinism."[33]

Whereas Hughes remains largely within a traditional historical framework, "social constructivists" take an avowedly more sociological view, arguing that what leads to "closure" regarding a "successful" technological design has more to do with the interest groups surrounding an artifact than it does technical elements of the device itself. Thus, Pinch and Bijker argue that what defines the "success" of the modern "safety" bicycle over the penny-farthing with the high front wheel has to do with the way interested parties, including women riders, macho male racers, and bicycle engineers, reached "closure" in their thinking about issues of safety, speed, and vibration. Thus, the social construction of technology (SCOT) focuses on the negotiations among the network of interested parties, constituting what Bijker elsewhere refers to as a "technological frame" or system of thought and practice in which the device is embedded.[34]

Taking the constructivist argument one step further are the "network theorists" typified by the work of Michel Callon, John Law, and Bruno Latour, mentioned previously in the context of the sociology of scientific knowledge. Here the key concept is that of the "actor network"—a group of entities that includes not only people but also theories, technical devices, political institutions and policies, even the natural environment. Together this network of animate "actors" and inanimate "actants" constitutes a seamless "web." These "heterogeneous elements" are all equally important and must be considered "symmetrically," that is, as equally important. When sufficient actors have been "enrolled" or brought together in support of a sociotechnical system, closure is accomplished and a "black box" now exists. Thus, John Law argues that the "success" of fifteenth-century Portuguese voyages of exploration along the African coast was due not merely to technical changes in vessel type but also to the combined availability of new navigational instruments and maritime tables and charts along with the "invention" or understanding of how to use the *volta*, a circular pattern of prevailing winds and currents. Together, and only together, did these equally important heterogeneous elements combine in a stabilized network, and thereby allow the Portuguese to sail out of Lisbon, track along the African coast out of sight of land, and, more importantly, return again safely. In this way a sociotechnical problem is defined and an acceptable solution found.[35]

In spite of the distinctions between these views, there is a strong common thread running through them, which has led to them being thought of collectively as "social constructivism." As should be clear from the brief case references noted, the constructivist approach tends to look at technology within the framework of systems or networks in which "societal" components shape or "construct" the technical outcome, which in turn, of course, can influence cultural and institutional values. Here the emphasis is on human choice and contingency rather than linearly deterministic technological change. In contrast to SCOT, which tends to privilege the social or human element in its construction of technical systems, actor network theory, as the Portuguese sailing example suggests, does not. Indeed, it tends to blur, if not deny, any meaningful distinction between the living "socio-" and the nonliving "technical." In part it is this conflation which has lead to significant controversy regarding the latter's descriptive accuracy and worth.[36] Nonetheless, by roughly the mid- to late 1980s, a generally constructivist mode of analysis utilizing empirical case studies, often historical in nature, had come to typify much sociological study of technology.

Philosophy of Technology

Prior to both historians and sociologists coming to recognize technology as a worthy subject, philosophers had already begun to explore the implications of technology for their own areas of study. As with many of the other related disciplines, philosophy of technology as a conscious and distinct concern traces its origins to the late nineteenth century, when the German philosopher, Ernst Kapp coined the phrase "philosophy of technology" in his 1877 book, *Grundlinien einer Philosophie der Technik* [Outlines of a Philosophy of Technology]. Kapp viewed technology in the form of tools and weapons as "organ projections," which he believed were deserving of philosophical reflection. Kapp and other engineering-trained or -oriented philosophers, such as the turn-of-the-century Russian Peter K. Engelmeier, and the mid-twentieth-century German Friedrich Dessauer, represent a strain of philosophical inquiry that Carl Mitcham has called "engineering philosophy of technology." Here the emphasis is on "philosophy of *technology*" rather than on "*philosophy* of technology," that is, a technologically oriented philosophy in contrast to ways of thinking *about* technology. In general such individuals held positive views of technology; Engelmeier, for example, sought to apply engineering rationality throughout society and supported the technocracy movement during the 1920s. Dessauer, an engineer-entrepreneur who did pioneering work in the development of X-ray technology, was particularly interested in the process of technological creation. He viewed modern engineering as a new way to live in the world, one that had a transcendent, almost moral and religious, overtone to it. More recently the Argentine-Canadian Mario Bunge has focused on what he calls "technophilosophy," an attempt to explain reality along technoscientific lines. In Bunge's mind, the solution to societal problems lies not in less but rather more technology; thus, he is sharply critical of postmodern constructivist views of technology. Taken together then, this strain within the philosophy of technology tends to favor or privilege technical criteria and paradigms over humanistic concerns.[37]

In contrast to "engineering philosophy of technology," a more contextual "humanities" philosophy that elevates the nontechnical dimension has come to characterize the field over the past several decades. This so-called humanities philosophy of technology also has a long tradition that includes the more historical and sociological, but still philosophically important, writers Lewis Mumford and Jacques Ellul, and the more professional philosophers José Ortega y Gasset and Martin Heidegger. All of these thinkers sought, on the one hand, to characterize technology by distinguishing different levels or types so that, on the other, they could identify the sorts of technology that give primacy to human beings over technical concerns per se. Thus, Mumford's authoritarian "megamachine" is out of sync with a more human-scaled "biotechnics," just as Ellul sees the "technical phenomenon"—the all-encompassing "technique" of modern technological society—as going beyond traditional "technical operation." In his turn, Ortega criticized our tendency to employ modern technical capabilities to accomplish ends that we had not yet consciously considered in any serious humanistic way, while Heidegger argued for a deeper "questioning" concerning technology's essence or nature. Heidegger, like Mumford and Ellul, wants to distinguish between more traditional technologies, those tied more directly to the earth, for example, wind- and water mills, with those modern technologies, such as coal fired power plants, that transform nature into artifacts solely for purposes of human consumption without thought for spirituality. It was precisely such calls as these that coalesced in the 1960s and 1970s into a newly "self-conscious" burgeoning of philosophical interest in technological issues among professional scholars.[38]

This self-conscious concern was reflected in a series of academic conferences, new journals and publications, and eventually the establishment of a professional society. Beginning as early as 1962, the Center for the Study of Democratic Institutions, which employed John Wilkinson to translate and would subsequently publish Ellul's *The Technological Society*, hosted a conference on "The Technological Order." This was followed by a 1965 symposium on philosophy of technology at the annual Society for the History of Technology meeting, a 1973 international symposium on history and philosophy of technology at the University of Illinois at Chicago, and conferences in 1975 an 1977 at the University of Delaware organized by Paul Durbin, who would go on to establish and edit both the *Philosophy and Technology Newsletter* and an annual series, *Research in Philosophy and Technology*. Out of these conferences and publications finally emerged in 1980–1981 the establishment of the Society for Philosophy and Technology, wherein the "and" (as opposed to an "of") reflected a desire to be as all-inclusive as possible.[39]

As has been the case with sociology of science and of technology, much of the best thinking in philosophy of technology has occurred within the journal literature. Nonetheless, several important book-length works reflected the humanistic or societal concerns regarding technology as they emerged in the 1970s. Among them must be counted the work of the political philosopher Langdon Winner, whose *Autonomous Technology* (1977) examined and deepened our understanding of the idea of a technological imperative, first raised by Ellul. Without succumbing to a simplistic determinism, Winner further examined the theme of technology's inherently political nature in a series of essays eventually collected as *The Whale and the Reactor* (1986). One of Winner's most important themes is that technologies become "forms of life" so deeply embedded within society that, in effect, they create new worlds in which existence without these technologies—electricity, automobiles, telephones, computers—would be almost unthinkable. Related to this theme, he further maintains that all technologies, or artifacts, have "politics." That is, all artifacts by their very design and nature express relationships of power and authority in some way or another. Ultimately Winner calls for society to pay far more attention to the goals we should be setting for ourselves vis-à-vis our technology. In short, we need to wake up from our "technological somnambulism."[40]

Another important figure, although seemingly less immediately concerned with the social consequences of technology, is the American phenomenologist Don Ihde, who has written an important series of monographs focusing on the inherently technological nature of human beings and the ways technology, which due to its inherently value-laden nature is always "non-neutral," extends or mediates human experience within the world.[41]

Finally in this period from the late 1970s to the mid-1980s, mention should be made of the work of Albert Borgmann, especially his *Technology and the Character of Contemporary Life* (1984). Borgmann distinguishes between what he calls "focal things" (e.g., a fireplace) and technological "devices" (e.g., a central heating system). Although both provide heat, in Borgmann's mind the first "is inseparable from its context," while the second tends to split ends and means. In other words, "the relatedness of the world is replaced by . . . machinery," and technology comes to dominate nature. Only by attending to focal things will we be able to solve society's problems. Here, as in Winner's work, we find expressed an increased sensitivity to the contextual nature of technology and a plea to society to consider more fully the ends toward which it directs that technology.[42]

In addition to these fairly broad ranging studies, philosophers also addressed a number of more specific technological issues and questions in their work, and in particular began to investigate the ethical issues associated with a number of key technoscientific developments. Unfortunately, this is not an area that has been as fully integrated into STS studies as broadly as it might be.[43] Perhaps the most general analysis of ethical responsibility was presented by Hans Jonas in his 1984 study, *The Imperative of Responsibility*, in which he argued that "the new kinds and dimensions of [technological] action require a commensurate ethic of foresight and responsibility." Also insightful is the work of the theologian-ethicist Ian Barbour, especially his *Ethics in an Age of Technology* (1992). Other more specific areas of applied ethical concern have included the nuclear, environmental, biomedical, computer, scientific research, and engineering fields. Representative examples of such an approach, which often includes work by nonphilosophically trained technical professionals, would include the following: Kristin Schrader-Frechette, *Nuclear Power and Public Policy* (1980); Avner Cohen and Steven Lee, eds., *Nuclear Weapons and the Future of Humanity* (1986); Eugene Hargrove, *Foundations of Environmental Ethics* (1989); Tristram H. Engelhardt Jr., *The Foundations of Bioethics* (1986); Deborah Johnson, *Computer Ethics* (1985); William Broad and Nicholas Wade, *Betrayers of the Truth* (1982); and Stephen H. Unger, *Controlling Technology: Ethics and the Responsible Engineer* (1982). As should be clear from these examples, by the mid- to late 1980s, philosophy of technology too had taken the societal issues associated with technology as a main focus of its attention.[44]

Taken together, all of these separate, largely disciplinary developments reflected an increased understanding of the complexities of modern science and technology in contemporary society. Each field sought to bring to bear its insights as to the nature of scientific and technological knowledge, and to explore both the obvious benefits, as well as previously ignored negative consequences. Even if not truly interdisciplinary, there were obvious interactions between fields with philosophers and especially sociologists utilizing historical case studies, while historians increasingly adopted constructivist metaphors in explaining the contextual development of science and technology. Thus by the mid-1980s, the broader STS field and its disciplinary-oriented components had achieved a certain level of maturity, so to speak. In the next chapter I turn to a description of what the field looks like now, including a brief pause to discuss the nature of interdisciplinarity and what that means for STS as a field of study.

Notes

1. Thomas S. Kuhn, *The Structure of Scientific Revolutions* (Chicago: University of Chicago Press, 1962, 2d ed., 1970). It is interesting to note, and reflective of his influence, that constructivist philosophers, historians, and sociologists of science have all traced and laid heritage claims to Kuhn. Although trained originally as a physicist, Kuhn made his intellectual contribution by writing about the history of science—for example, studies on the Copernican revolution in astronomy and on black-body theory—while at the same time treating philosophical questions, especially in his work on the concept of scientific revolutions. Sociologists of science, as that field moved away from traditional positivist-oriented debates regarding scientific knowledge, quickly recognized a kindred soul in Kuhn, or at least in the sociological implications of his work. Somewhat ironically, given his influence on these three fields that have coalesced at one level to create the current interest in science studies, Kuhn himself was apparently never particularly at home in any one of the professional associations of any of the specific disciplines, or at least this is the suggestion of Ronald Giere in "Kuhn's Legacy for North American Philosophy of Science," one of a linked series of obituaries in *Social Studies of Science* 27 (June 1997): 496–98. Kuhn's other works include: *The Copernican Revolution: Planetary Astronomy in the Development of Western Thought* (Cambridge: Harvard University Press, 1957); *The Essential Tension: Selected Studies in Scientific Tradition and Change* (Chicago: University of Chicago Press, 1979); and *Black-Body Theory and the Quantum Discontinuity, 1894–1912* (Chicago: University of Chicago Press, 1987).

2. Melvin Kranzberg, "Kranzberg's Laws," *Technology and Culture* 28 (July 1986): 544–60, and reprinted in *In Context: History and the History of Technology—Essays in Honor of Melvin Kranzberg*, ed. Stephen H. Cutcliffe and Robert C. Post (Bethlehem, Pa.: Lehigh University Press, 1989), 244–58.

3. Among the leading positivists was Rudolf Carnap; see his *Philosophical Foundations of Physics* (New York: Basic Books, 1966) and *Introduction to the Philosophy of Science* (New York: Basic Books, 1966).

4. Norwood Russell Hanson, *Patterns of Discovery: An Inquiry into the Conceptual Foundations of Science* (Cambridge: Cambridge University Press, 1968), 6.

5. Robert Klee, *Introduction to the Philosophy of Science: Cutting Nature at Its Seams* (Oxford: Oxford University Press, 1997) provides a recent and useful summary of the positivist school and the challenges to it that have subsequently emerged; see especially chapters 2–4.

6. Kuhn, *The Structure of Scientific Revolutions*, 24.

7. Kuhn, *The Structure of Scientific Revolutions*, 103.

8. Kuhn, *The Structure of Scientific Revolutions*, 170.

9. Klee, *Introduction to Philosophy of Science*, especially chapter 7, offers a useful summary of Kuhn from the perspective of a philosopher who wishes to hold on to a realist position. Klee also discusses the antirealist position, especially through an analysis of the philosopher of physics Bas van Fraassen, 216–18, 226–31. Among Van Fraassen's works are *The Scientific Image* (Oxford: Oxford University Press, 1980); *An Introduction to the Philosophy of Time and Space* (New York: Columbia University Press, 1985); and *Laws and Symmetry* (Oxford: Oxford University Press, 1989).

10. Both Andrew Webster, *Science, Technology, and Society* (New Brunswick, N.J.: Rutgers University Press, 1991) and Randall Collins and Sal Restivo, "Development, Diversity, and Conflict in the Sociology of Science," *The Sociological Quarterly* 24 (Spring 1983): 185–200 discuss the emergence of the sociology of science as a field of study. For Merton, see his *Social Theory and Social Structure* (New York: Free Press, 1949) and his later collection of essays, *The Sociology of Science* (Chicago: University of Chicago Press, 1973).

11. Representative of Mulkay's early work are his article, "Norms and Ideology in Science," *Social Science Information* 15 (1976): 637–56, and his book-length study, *Science and the Sociology of Knowledge* (London: George Allen & Unwin, 1979).

12. David Bloor, *Knowledge and Social Imagery*, 2d ed. (Chicago: University of Chicago Press, 1991). Also see Bloor's recent essay, "Remember the Strong Program?" *Science, Technology, & Human Values* 22 (Summer 1997): 373–85 and the coauthored book by Barry Barnes, David Bloor, and John Henry, *Scientific Knowledge: A Sociological Analysis* (Chicago: University of Chicago Press, 1996).

13. See, for example, Harry Collins, *Changing Order: Replication and Induction in Scientific Knowledge* (Chicago: University of Chicago Press, 1985, 2d ed., 1992) and Trevor Pinch, *Confronting Nature: The Sociology of Solar-Neutrino Detection* (Dordrecht: Reidel, 1986). These two scholars also jointly authored *The Golem: What Everyone Should Know about Science*, 2d ed. (Cambridge: Cambridge University Press, 1998), which summarizes for a general audience these and other empirical studies from a similar perspective. Karl Popper argued that only after the failure of deliberate, scientifically rigorous attempts to "falsify" a theory could it become acceptable as true. See, for example, *The Logic of Scientific Discovery* (New York: Basic Books, 1959) or his more accessible *Objective Knowledge: An Evolutionary Approach* (Oxford: Oxford University Press, 1972).

14. On self-reflexivity, see Steven Woolgar, ed., *Knowledge and Reflexivity: New Frontiers in the Sociology of Knowledge* (Beverly Hills, Calif.: Sage, 1988) and *Science: The Very Idea* (London: Tavistock, 1988). Also valuable in this regard are Karin Knorr-Cetina and Michael Mulkay, eds., *Science Observed: New Perspectives on the Social Study of Science* (Beverly Hills, Calif.: Sage, 1983) and G. Nigel Gilbert and Michael Mulkay, *Opening Pandora's Box: A Sociological Analysis of Scientists' Discourse* (Cambridge: Cambridge University Press, 1984).

15. Bruno Latour and Steve Woolgar, *Laboratory Life: The Construction of Scientific Facts*, 2d ed. (Princeton: Princeton University Press, 1986); Bruno Latour,

Science in Action: How to Follow Scientists and Engineers through Society (Cambridge: Harvard University Press, 1987), see especially 13, 15, and quotation, 258. In *Science in Action*, Latour includes the study of technology in addition to science per se, in what he calls "technoscience." Prior to this point, most sociologists focused primarily on the formation of scientific knowledge only. Both Klee in his *Introduction to Philosophy of Science*, from a philosophical stance, and Paul Gross and Norman Levitt in *Higher Superstition: The Academic Left and Its Quarrels with Science* (Baltimore: Johns Hopkins University Press, 1994), from the perspective of science, take issue with what they see as the antirealist position that Latour represents. In his most recent book, *Pandora's Hope: Essays on the Reality of Science Studies* (Cambridge: Harvard University Press, 1999), Latour has backed away from some of the antirealist implications of his earlier work. Latour, with training in anthropology and philosophy, is generally referred to as a sociologist, although there is debate in some quarters as to the appropriateness of such an appellation, given his general unwillingness to privilege the human over the nonhuman in terms of explaining technoscience.

16. See Jerry Gaston, "Sociology of Science and Technology," in *A Guide to the Culture of Science, Technology, and Medicine*, ed. Paul T. Durbin (New York: Free Press, 1980, 1984), 468–70, for a brief discussion of the founding of 4S, but also see Merton's description of the institutionalization of the sociology of science, "The Sociology of Science: An Episodic Memoir," in *The Sociology of Science in Europe*, ed. Merton and Jerry Gaston (Carbondale: Southern Illinois University Press, 1977), 3–141, in which he notes the increased public concern regarding the societally "problematical" nature of science (p. 112). For representative samples of the work of Bernard Barber, see *Science and the Social Order* (New York: Free Press, 1952) and of Joseph Ben-David, see *The Scientist's Role in Society: A Comparative Study* (Englewood Cliffs, N.J.: Prentice Hall, 1971).

17. William Whewell, *History of the Inductive Sciences*, 3 vols. Reprint ed. (London: Cass, 1967). Whewell also published a two-volume *Philosophy of the Inductive Sciences*, 1840. Reprint ed. (London: Cass, 1967).

18. See Arnold Thackray, "History of Science," especially 8–12, in Durbin, ed., *Guide*, for a brief summary of nineteenth-century precursors to Sarton.

19. See Thackray, "History of Science," 14–15 for a brief discussion of this period. For specific examples of the work of J. D. Bernal, see *The Social Function of Science* (London: Routledge, 1939), but also his later *Science in History*, 4 vols. (London: Watts, 1954), which surveys the social context of science. Max Weber's earlier 1904–1905 study of the relationship between Protestantism and the industrial revolution was translated into English in 1930 and subsequently had widespread influence, *The Protestant Ethic and the Spirit of Capitalism*, trans. Talcott Parsons (Cambridge: Harvard University Press, 1930), but also see his 1920 essay "Science as a Vocation," in *From Max Weber: Essays in Sociology*, ed. H. H. Gerth and C. W. Mills, reprint ed. (New York: Oxford University Press, 1946). Joseph Needham's multivolume work, *Science and Civilization in China* (Cambridge: Cambridge University Press, 1954–), coauthored volumes of which continue to appear, even after his death, has shown the ongoing influence of this perspective.

20. See, for example, Koyré's *From the Closed World to the Infinite Universe* (Baltimore: Johns Hopkins University Press, 1956) which treats the sixteenth-

and seventeenth-century shift from a geocentric religious world view to a heliocentric and scientific cosmological view.

21. I. Bernard Cohen, *The Birth of a New Physics* (Garden City, N.J.: Doubleday, 1969); Thomas S. Kuhn, *The Copernican Revolution: Planetary Astronomy in the Development of Western Thought* (Cambridge: Harvard University Press, 1957); Henry Guerlac, *Lavoisier: The Crucial Year* (Ithaca, N.Y.: Cornell University Press, 1961); and Charles C. Gillispie, *Lazare Carnot, Savant* (Princeton: Princeton University Press, 1971).

22. Jerome J. Ravetz, *Scientific Knowledge and Its Social Problems* (Oxford: Clarendon Press, 1971); Gerald Holton, *Thematic Origins of Scientific Thought: Kepler to Einstein* (Cambridge: Harvard University Press, 1973); Mary B. Hesse, *The Structure of Scientific Inference* (London: Macmillan, 1974); Charles E. Rosenberg, *No Other Gods: Science and American Social Thought* (Baltimore: Johns Hopkins University Press, 1976); and Daniel J. Kevles, *The Physicists: The History of a Scientific Community in Modern America* (New York: Knopf, 1978).

23. David O. Edge and Michael J. Mulkay, *Astronomy Transformed: The Emergence of Radio Astronomy in Britain* (New York: John Wiley, 1976) and Steven Shapin and Simon Schaffer, *Leviathan and the Air-Pump: Hobbes, Boyle, and the Experimental Life* (Princeton: Princeton University Press, 1985).

24. Schaffer, *Leviathan and the Air-Pump*, 344. For a realist critique of Shapin and Schaffer's argument, see Klee, *Introduction to Philosophy of Science*, 74–79. Shapin has expanded upon his ideas regarding the roles of "civility" and "trust" in accounting for the establishment of scientific "truth" among seventeenth-century English gentlemen philosopher-scientists in *A Social History of Truth: Civility and Science in Seventeenth-Century England* (Chicago: University of Chicago Press, 1994).

25. Samuel Smiles, *Lives of the Engineers, with an Account of their Principal Works: Comprising Also a History of Inland Communication in Britain* (London: John Murray, 1862), but see also his *Industrial Biography: Iron-workers and Toolmakers* (London: John Murray, 1863). Also useful is Thomas P. Hughes, ed., *Selections from Lives of the Engineers* (Cambridge: MIT Press, 1966).

26. Charles Singer et al., eds., *A History of Technology*, 8 vols. (Oxford: Oxford University Press, 1955-84). The original five volumes of this survey of technology from the ancient world to 1900 were published in the mid-1950s, with subsequent twentieth-century volumes appearing thereafter. Singer also wrote surveys in the history of science and medicine, including *A Short History of Scientific Ideas to 1900* (Oxford: Oxford University Press, 1959) and, with Ashworth E. Underwood, *Short History of Medicine* (Oxford University Press, 1962).

27. Lewis Mumford, *Technics and Civilization* (New York: Harcourt, Brace and Company, 1934), quotation, 6; and Siegfried Giedion, *Mechanization Takes Command: A Contribution to Anonymous History* (New York: Oxford University Press, 1948). Arthur P. Molella, "The First Generation: Usher, Mumford, and Giedion," in *In Context*, ed. Cutcliffe and Post (Bethlehem, Pa.: Lehigh University Press, 1989), 244–58, 88–105, offers an excellent introduction to the work of Mumford and Giedion as early historians of technology who took a contextual approach to their studies. Molella also includes Abbott Payson Usher, an economic historian, who sought to include the human element in explaining the process of technological innovation, but he notes that Usher's work, due to its

largely internalist focus, is more theoretically suggestive than primarily social in its historical analysis. See, for example, Usher's *A History of Mechanical Inventions* (New York: McGraw-Hill, 1929; rev. ed. Cambridge: Harvard University Press, 1954).

28. On the foundation of SHOT, see John Staudenmaier, *Technology's Storytellers: Reweaving the Human Fabric* (Cambridge: MIT Press, 1985), especially chapter 1. Also useful is Carroll W. Pursell Jr., "History of Technology," especially 73–74, in *Guide*, ed. Durbin. SHOT's original statement of purpose, as contained in *Technology and Culture*, has been somewhat revised of late to include "the relations of technology to politics, economics, labor, business, the environment, public policy, science, and the arts." Nonetheless, the contextual intent is clear and continuous.

29. Lynn White jr., *Medieval Technology and Social Change* (Oxford: Oxford University Press, 1962); Ruth Schwartz Cowan, "The 'Industrial Revolution' in the Home: Household Technology and Social Change in the 20th Century," *Technology and Culture* 17 (January 1976): 1–23 and *More Work for Mother: The Ironies of Household Technology from the Open Hearth to the Microwave* (New York: Basic Books, 1983); Thomas P. Hughes, *Networks of Power: Electrification in Western Society, 1880–1930* (Baltimore: Johns Hopkins University Press, 1983). Merritt Roe Smith and Steven C. Reber, "Contextual Contrasts: Recent Trends in the History of Technology," in *In Context*, ed. Cutcliffe and Post, 133–49, contains an excellent summary of four works, including Cowan and Hughes, that reflect the range of contextual approaches within the history of technology.

30. William F. Ogburn, *Social Change with Respect to Culture and Original Nature* (New York: Viking Press, 1922); Ogburn with Jean L. Adams and S. C. Gilfillan, *The Social Effects of Aviation* (Boston: Houghton Mifflin, 1946); Ogburn and Meyer F. Nimkoff, *Technology and the Changing Family* (Boston: Houghton Mifflin, 1955); Sean Colum Gilfillan, *The Sociology of Invention* (Chicago: Follett, 1935); and Gilfillan, *Invention and the Patent System* (Washington, D.C.: Joint Economic Committee of Congress, 1964). Ron Westrum, *Technologies and Society: The Shaping of People and Things* (Belmont, Calif.: Wadsworth, 1991), chapter 3, contains a nice summary of the Ogburn generation of sociologists of technology. Westrum notes, somewhat ironically, that in part it may well have been the very success of the emerging interest in the history of technology that helped to undercut the sociology of technology. In fact, Ogburn was elected as SHOT's first president.

31. Georges Friedmann, *Industrial Society: The Emergence of the Human Problems of Automation*, trans. by H. L. Sheppard (New York: Free Press, 1955) and *Anatomy of Work: Labor, Leisure, and the Implications of Automation*, trans. by W. Rawson (New York: Free Press, 1962); Jacques Ellul, *The Technological Society*, trans. by John Wilkinson (New York: Knopf, 1964); Ivan Illich, *Tools for Conviviality* (New York: Harper and Row, 1973).

32. Donald MacKenzie and Judy Wajcman, eds., *The Social Shaping of Technology: How the Refrigerator Got Its Hum* (Milton Keynes, U.K.: Open University Press, 1985), 2; Wiebe E. Bijker, Thomas P. Hughes, and Trevor Pinch, eds., *The Social Construction of Technological Systems: New Directions in the Sociology and History of Technology* (Cambridge: MIT Press, 1987), 14. For an overview of this

turn to technology, see Steve Woolgar, "The Turn to Technology in Social Studies of Science," *Science, Technology, & Human Values* 16 (Winter 1991): 20–50.

33. For Hughes's ideas on systems theory, see *Networks of Power*; "The Evolution of Large Technological Systems," in *Social Construction of Technological Systems*, ed. Bijker et al., 51–82; and *American Genesis: A Century of Invention and Technological Enthusiasm, 1870–1970* (New York: Viking, 1989), where he expands further upon these ideas.

34. Trevor I. Pinch and Wiebe E. Bijker, "The Social Construction of Facts and Artifacts: Or How the Sociology of Science and the Sociology of Technology Might Benefit Each Other," and Wiebe E. Bijker, "The Social Construction of Bakelite: Toward a Theory of Invention," in *Social Construction of Technological Systems*, ed. Bijker et al., 17–50, 159–87. See also Bijker's more recent study, *Of Bicycles, Bakelites, and Bulbs: Toward a Theory of Sociotechnical Change* (Cambridge: MIT Press, 1995), in which he expands upon these ideas.

35. John Law, "Technology and Heterogeneous Engineering: The Case of Portuguese Expansion," in *Social Construction of Technological Systems*, ed. Bijker et al., 111–34. See also Michel Callon's study of electric vehicles in France, "Society in the Making: The Study of Technology as a Tool for Sociological Analysis," in *Social Construction of Technological Systems*, ed. Bijker et al., 83–103.

36. Among those who have critiqued social constructivism are Langdon Winner in an essay entitled "Upon Opening the Black Box and Finding It Empty: Social Constructivism and the Philosophy of Technology," *Science, Technology, & Human Values* 18 (Summer 1993): 362–78. Winner recognizes certain valuable contributions of constructivism but finds it lacking in terms of its "almost total disregard for the social consequences of technical choice," p. 368, and its general failure to "offer . . . judgement on what it all means," p. 375. Representative of those realists, often practicing scientists who distrust or deny the antirealist implications of constructivism, especially that of SSK, is the work of Gross and Levitt, *Higher Superstition* already noted, but also see Gross, Levitt, and Martin W. Lewis, eds., *The Flight from Science and Reason* (New York: New York Academy of Sciences, 1996).

37. Carl Mitcham's wide-ranging survey of philosophy of technology, *Thinking through Technology: The Path between Engineering and Philosophy* (Chicago: University of Chicago Press, 1994) provides a bibliographic tour de force of the field, in which Mitcham divides the literature between engineering and humanities approaches. Ernst Kapp, *Grundlinien einer Philosophie der Technik*, reprint ed. (Dusseldorf: Stern-Verlag Janssen, 1978); Peter K. Engelmeier, *Filosofia techniki*, 4 vols. (Moscow, 1912); Friedrich Dessauer, *Philosophie der Technik: Der Problem der Realisierung* (Bonn: F. Cohen, 1927) and *Streit um die Technik* (Frankfurt: J. Knecht, 1956); and Mario Bunge, see, for example, his "Toward a Philosophy of Technology," in *Philosophy and Technology*, ed. Carl Mitcham and Robert Mackey (New York: Free Press, 1972), 62–76, all reflect an internalist engineering-oriented philosophy that views the technological perspective as a paradigmatic way for living in the world.

38. See, for example, Lewis Mumford, *Myth of the Machine*, 2 vols. (New York: Harcourt Brace Jovanovich, 1967, 1970); Ellul, *The Technological Society*, especially 19–22; José Ortega y Gasset, "Thoughts on Technology," in *Philosophy*

and Technology, ed. Mitcham and Mackey, 290–313; and Martin Heidegger, "The Question Concerning Technology," (1954), in *The Question Concerning Technology and Other Essays* (New York: Harper and Row, 1977). Mitcham, *Thinking through Technology*, chapter 2, contains a good discussion of this group of philosophers. It should be noted that the boundaries between the engineering and humanities approaches to philosophy of technology are not as clearly defined as suggested here or by Mitcham, with individuals frequently crossing over lines on the map. Nonetheless, the distinction is a useful heuristic for guiding a reader through the literature.

39. See Mitcham, *Thinking through Technology*, 9–11, for a more detailed account of this institutional development. It is, however, instructive to note two things. Interestingly, 1962, the year of the "Technological Order" conference, was also the same year in which Kuhn published *The Structure of Scientific Revolutions* and in which Rachel Carson's *Silent Spring* also appeared. Secondly, it was no accident that Melvin Kranzberg, having recently established SHOT as an organization with interests distinct from those of the historians of science, should have extended an open welcome to the philosophers of technology, and subsequently published an issue of *Technology and Culture* (Summer 1966) based on the previous year's symposium. He subsequently published as a supplement to *Technology and Culture* 14 (April 1973) a preliminary version of Mitcham and Mackey's *Bibliography of the Philosophy of Technology* (Chicago: University of Chicago Press, 1973) and played a role in the 1975 Chicago "History and Philosophy of Technology Conference."

40. Langdon Winner, *Autonomous Technology: Technics-out-of-Control as a Theme in Political Thought* (Cambridge: MIT Press, 1977) and *The Whale and the Reactor: A Search for Limits in an Age of High Technology* (Chicago: University of Chicago Press, 1986).

41. Don Ihde, *Technics and Praxis: A Philosophy of Technology* (Boston: D. Reidel, 1979); *Existential Technics* (Albany: State University of New York Press, 1983); and *Technology and the Lifeworld: From Garden to Earth* (Bloomington: Indiana University Press, 1990).

42. Albert Borgmann, *Technology and the Character of Contemporary Life* (Chicago: University of Chicago Press, 1984), 42, 47. See also the latter's more recent update and extension of his ideas, *Crossing the Postmodern Divide* (Chicago: University of Chicago Press, 1992) and his even newer study, *Holding on to Reality: The Nature of Information at the Turn of the Century*. (Chicago: University of Chicago Press).

43. On applied ethical studies associated with technoscience, see Carl Mitcham, "Etica Sobre y Dentro Ciencia y Tecnología," *Theoria* 14 (Fall 1999).

44. Hans Jonas, *The Imperative of Responsibility: In Search of an Ethics for the Technological Age* (Chicago: University of Chicago Press, 1984); Ian Barbour, *Ethics in an Age of Technology* (New York: HarperCollins, 1992); Kristin Schrader-Frechette, *Nuclear Power and Public Policy: The Social and Ethical Problems of Fission Technology*, rev. ed. (Boston: D. Reidel, 1983); Avner Cohen and Steven Lee, eds., *Nuclear Weapons and the Future of Humanity: The Fundamental Questions*, Philosophy and Society Series (Savage, Md.: Rowman & Littlefield, 1986); Eugene Hargrove, *Foundations of Environmental Ethics* (Englewood Cliffs, N.J.: Prentice Hall, 1989); Tristram H. Engelhart Jr., *The Foundations of*

Bioethics (Oxford: Oxford University Press, 1986); Deborah Johnson, *Computer Ethics* (Englewood Cliffs, N.J.: Prentice Hall, 1985); William Broad and Nicholas Wade, *Betrayers of the Truth* (New York: Simon and Schuster, 1982); and Stephen H. Unger, *Controlling Technology: Ethics and the Responsible Engineer*, 2d ed. (New York: Wiley, 1993). For a more extended bibliographical listing of philosophical works in these areas, see Mitcham, *Thinking through Technology*, but see also Mitcham and Leonard J. Waks, "Technology in Applied Ethics: Moving from the Margins to the Center," *Bulletin of Science, Technology & Society* 16 (1996): 217–26.

3

Interdisciplinarity and the Current State of STS

The question is not how to eliminate cultural values but instead how to find out which cultural values structure science and whether different or better sciences would result if other values replaced the ones currently in place.

— David Hess, *Science Studies*

As suggested by the previous chapter's outline of disciplinary developments in the socially oriented study of science and technology, STS by roughly the mid-to-late 1980s had matured as a multidisciplinary, if not interdisciplinary, field of study. It had developed a certain body of scholarship that had become widely read and cited, and it had established itself institutionally in terms of academic programs and professional organizations, topics that will be treated in detail in chapter 4. The question remains as to the current state of affairs within STS, which will be the main focus of this chapter. Before proceeding to summarize where scholarship in the field appears to be today, however, it is first necessary to take a brief detour to discuss the issue of interdisciplinarity and what that might mean for a field such as STS.

The Issue of Interdisciplinarity

Interdisciplinarity is a term frequently used to convey a range of often overlapping definitions. Julie Thompson Klein, in her book *Interdisciplinarity: History, Theory, and Practice*, suggests a distinction between interdisciplinary research and interdisciplinary studies education. In the area of science and technology, the former could be represented by such fields as biochemistry and materials science, while the latter might well include STS studies. She also suggests a continuum running from multidisciplinarity through interdisciplinarity to transdisciplinarity. In this view, multidisciplinarity involves the juxtaposition of more traditional disciplines in order to solve some problem or to gain insight, while interdisciplinarity involves more extensive integration involving borrowing, common problem solving, increased consistency of subject matter or methodology, or even the emergence of an interdiscipline. Transdisciplinarity, in contrast, goes beyond either of the first two to create conceptual frameworks capable of influencing more than one discipline, one example being general systems theory. Another might well be postmodernism. Although Klein cautions that there is "no inevitable progression from 'multidisciplinarity' through 'interdisciplinarity' to 'transdisciplinarity,'" STS, which could be said to have begun in "multidisciplinarity," I would argue has progressed to a level of "interdisciplinarity" and, in some quarters, is perhaps striving to achieve some sense of "transdisciplinarity."[1]

Taking a somewhat different tack, sociologist Gary Bowden argues that while "STS, as a scholarly pursuit, has come of age," it is still largely a topically focused, multidisciplinary field in which disciplinary master narratives still dominate. He suggests that there are three main explanatory methods within STS—topic focused, issue focused, and combined—and argues that choice of method largely depends on how one defines the field. Topic-focused research, the most common approach within STS, specifies the subject, which is then examined from the particular disciplinary perspective of the scholar. Here it is the subject matter of science and technology, rather than any single methodology unique to STS, that holds the field together. Thus, a multiplicity of disciplinary perspectives will have something to offer by way of insight, but no one discipline is sufficiently illuminating by itself. A combined or interdisciplinary approach to the field would assume some particular or special characteristics to the subject matter, which in turn would require a unique methodology. As one of the main insights from STS is the societally contextual nature of scientific knowledge and technology, which undercuts their uniqueness, Bowden argues, without denying the future possibility of an integrative disciplinary focus, that combined methods have yet to prove themselves. He then goes on to argue that analytic, issue-focused methods, such as might be represented by postmodernism or the idea of reflexivity, are not linked necessarily to the substantive issue, in this case, of science and technology. Rather, they are approaches that transcend the disciplines, and hence in Bowden's mind they are "incompatible with STS." Thus, social studies *in* science and technology, such as represented by methodologically proper "reflexivity," should be viewed as falling outside the bounds of STS, in contrast to studies *of* science and technology as the substantive topic.[2]

Bowden goes on to argue that as STS has come of age, it has become "organized around the varying historical and cross-cultural manifestations of the relationship between social context and the processes, cultures, and institutions involved with understanding, manipulating, and using nature." In short, STS is "an amalgamation of [disciplinary rooted] contextualist approaches." Thus, "the main thread holding the fabric of STS together is the notion that S&T must be viewed *in context*." Given the sociohistorical contingent nature of this contextual/constructivist view of science and technology, Bowden believes it makes little sense to view science and technology as "unique" and hence there is no requirement for any sort of specialized scholarship. Only if STS were to focus on an issue such as science and technology's unique "status" within society, much as Women's Studies has done for women, might STS lay sufficient claim to "interdisciplinarity." Otherwise, it makes more sense to view STS, at least methodologically, as being closer to having "an adolescent identity" rather than "a stable adult personality."[3]

Henry Bauer, a chemist and STS scholar, has highlighted some of the barriers against interdisciplinarity due to the ways "disciplines differ in epistemology, in what they view as knowledge, and in opinion over what sort of knowledge is possible." Examples of interdisciplinary fields that do exist, such as biochemistry, are successful precisely because they entail "no major alterations of underlying epistemologies or methodologies." Nonetheless, he argues the value of STS because it brings a spectrum of valuable disciplinary perspectives to bear, no one of which is superior or inferior to another. The problem entails showing the value of both the practical applied side of STS problem solving along with theoretical understandings of science and technology. He highlights the epistemic difficulty of doing this, when "those who believe it *wrongheaded not to assume a discoverable reality* must work constructively with those to whom it is *wrongheaded to assume such discoverability*." While Bauer does suggest some useful insights from STS, such as the distinctions between science and technology, the fallibility of science, and the fact that science is a "social activity," he is not sanguine over their "axiomatic" acceptance by academia, let alone society as a whole.[4]

I would argue, somewhat in contrast to Bowden and Bauer, that the widely accepted usage across disciplines of such conceptualizations as contextualism and constructivism does, in fact, suggest that STS can be viewed in terms of an interdisciplinary effort along the lines suggested by Klein. In this regard, one possible way of thinking about the nature of STS is to think metaphorically in terms of an umbrella. When STS first emerged in the late 1960s and early 1970s, in the main it consisted of a group, one might even say hodge-podge, of disciplines somewhat randomly huddled together under the umbrella of STS. Interestingly, when the National Association of STS was created, it chose as its logo a stylized image of an umbrella, because it viewed itself as an umbrella organization covering a wide range of issues and perspectives. Included among the major disciplinary perspectives, as suggested in the previous chapter, were the history, philosophy, and sociology of science and technology. (There were of course others as well, including such fields as political science and anthropology.) Because many scientific and technical questions and issues are simply not analyzable on the grounds of single and separate disciplines, this multidisciplinary approach had distinct advantages. Moreover, STS has the related advantage of positive gains that can be had from transcending disciplinary boundaries, which are after all quite arbitrary, at least at certain levels. For example, abstraction-oriented philosophers of technology and case-study, detail-oriented historians of technology really have much to gain from being cognizant of and, where appropriate, utilizing each other's approach and methodologies, something that can be readily accomplished through an STS approach. Similarly, the interactions and borrowing of research approaches and terminology among the sociologists and historians represented in Wiebe Bijker et al.'s *The Social Construction of Technological Systems: New Directions in the Sociology and History of Technology*, which is usually pointed to as an early key work in the emergence of the SCOT (Social Construction of Technology) points to a certain level of interdisciplinarity. Within this collection, which admittedly emerged out of an interactive conference where such exchanges are more likely to take place, the authors frequently draw on and cross reference each other's work. In fact, the very title of the collection integrates the notion of "social construction" borrowed from SSK—the sociology of scientific knowledge—with the more historically based concept of "technological systems" usually associated with the work of Thomas Hughes. Thus the historian Hughes argues "the components of technological systems are socially constructed artifacts" and compares the "systems builders in their constructive activity" to John Law's "heterogeneous engineers," while sociologists Bijker, Pinch, Michel Callon, and John Law, in describing the "indissoluble link" between technology and society, draw upon the notion of a "seamless web" attributed to Hughes in his study of electric power generating systems, *Networks of Power*.[5]

It should perhaps be pointed out, as a not unimportant aside, that some scholars see such methodological sharing as something of a mixed blessing. For example, John Staudenmaier, in an extended essay review of Wiebe Bijker and John Law's edited volume *Shaping Technology/Building Society: Studies in Sociotechnical Change,* compares and contrasts the approaches of historians and sociologists of technology and asks how they can stimulate one another's work. At issue in this book, which includes contributions from both disciplinary perspectives, is how society reaches "closure" on a given technology given its underdetermined nature. Staudenmaier is interested in how sociologists and historians "generalize" from their case studies, including for the former the construction of vocabulary—for example, "closure"—relevant to the modeling of technological activity. He finds such neologisms "stimulating" historical tools when used as conceptual heuristics but "oppressive" when applied universally as "epistemological laws." Staudenmaier is also bothered when sociologists like Bruno Latour see "*only* actors" and thereby elect not to distinguish between the human and the nonhuman at work in technological construction, especially "if technological scholarship is going to call human decision makers to account for the morality, or even the aesthetics, of their choices."[6]

Having suggested the overlapping nature of somewhat arbitrarily defined disciplines, such as history, sociology, and philosophy, one can envision an umbrella, not under which these various related fields huddle, but rather one whose very fabric is constituted of panels, each of which represents a disciplinary approach, be it historical, sociological, or philosophical. Admittedly, Bowden might well still view this as a form of multidisciplinarity. Nevertheless, while the image is clearly inadequate in the sense that surely each discipline interacts with other fields of study beyond that which the two touching edges of an umbrella panel would suggest, at least all the panels come together at a point of sorts suggesting, even if not metaphorically perfect, the holism that must be there if the artifact is to function properly, or in this case if science and technology are to be understood fully.

It does appear that we have reached at least the point where scholars and educators have taken to borrowing methodology for purposes of common problem solving, which has been further marked by the emergence of scholarly journals, professional societies, formal graduate and undergraduate programs and departments, and even STS textbooks,[7] all of which will be discussed in further detail in the following chapter. Whether STS has reached a transdisciplinary level, one in which the fabric of our metaphorical umbrella might be truly "seamless," to borrow a turn of phrase from the lexicon of the sociologists of science and technology, however, is more open to debate.

The issue of interdisciplinarity can also be approached from another direction. At a 1989 conference at Cornell University convened to mark the twentieth anniversary of that institution's STS program, sociologist Dorothy Nelkin raised the following questions:

> STS is still struggling with a framework and a mission. Is its purpose to promote science, advance science, and frame policy that will advance scientific and technological development? Or is it a form of criticism focused on assessing and analyzing and critiquing science and technology decisions? Is it theoretical analysis of science as an extension of the sociology of knowledge and in effect a search for an understanding of science and society dimensions? Or is it a policy field intended to engineer solutions to such dilemmas and to design new policies?[8]

In many ways Nelkin's questions still remain unanswered, yet, in another way, as I suggest below and in the final chapter of this book, they have been answered, all in the affirmative. That is, STS has the potential, if viewed holistically, to be all of these things, which is not to say, however, that it can be all things to all people.

Analyzing and responding to the questions raised by Nelkin and during the meeting more generally, sociologist Susan Cozzens, then editor of *Science, Technology, & Human Values*, in a subsequent essay that has since been reprinted, conceptualized STS not as a field or academic discipline, but as a "movement." She views STS as having an "ideal," even if it does not as of yet have a theoretical core, "an integrated and accessible body of knowledge to inform [the movement]." She envisions and maps out STS as an "interdisciplinary network," albeit one whose members still have strong disciplinary ties. The answer for Cozzens is to complement this decentralized network with "a core [of] central questions or concepts that pull everyone together." In developing such a common core of problems and concepts, she argues, STS must "move away from specialization, toward broader relevance," a task she hopes will be accomplished through such interdisciplinary outlets as the Society for Social Studies of Science and its journal, *Science, Technology, & Human Values,* and professional meetings of associations such as the National Association for Science, Technology and Society (NASTS).[9]

Philosopher Paul Durbin, editor of *Philosophy & Technology*, in a follow-up article argues even more strongly that while STS may not yet have an authoritative body of theoretical work, what it does have is an "intellectual community." For Durbin, organizations such as NASTS and 4S, and their annual meetings, can provide the interdisciplinary locus within which to discuss the subjects, objects, aims, and methods of STS.[10]

More recently Wiebe Bijker has suggested that the "social construction" cases conducted during the decades of the 1980s and 1990s have provided enough empirical research that it is possible to take a "turn toward practice," a turn that would allow a more "democratic" public role in the shaping and control of technology.[11] In doing so he indirectly suggests the answer to each of Nelkin's questions is positive. Bijker argues that because "we live in a technological culture" we have "an obligation to try to *understand* [that] culture." At the same time he wants to *politicize* it, that is, "to make explicit the political dimensions of the role of science and technology, to question the self-evident character of technological culture, and to put science and technology on the public agenda for political deliberation." Finally, although he readily admits there will be differences of opinion, he hopes to *democratize* modern scientific and technological culture by engaging more citizens in such political deliberation. To assist in this process, Bijker suggests an extension of the standard constructivist model (SCOT), which tends to privilege the societal factors shaping technoscience, by means of what he calls the "technological frame" that helps structure the interactions among the "actors" of a relevant social group. Thus, the frame of existing practice guides future practice, not in a deterministic manner, but rather in terms of how technology, which can be "obdurate, hard, and . . . fixed," influences subsequent interactions and shapes culture. What we end up with then, in Bijker's terminology, is a "sociotechnical ensemble," in which society and technoscience are two sides of the same "sociotechnical coin."[12] In this way Bijker hopes to transcend both disciplinary boundaries and hard and fast distinctions between artifacts and society.

Two other metaphorical images may help further illustrate what Bijker has in mind here. One might think of those elements that go into the making of cloth or, perhaps even more homogeneously, a cake. In each case, distinct ingredients, whether they be the warp and woof threads that constitute a finely woven piece of tapestry or the flour, sugar, and eggs that constitute a cake, when combined in proper fashion result in a "seamless web" wherein the heterogeneous elements can no longer be readily distinguished. Thus, one can talk about both the distinctions among the threads or the ingredients utilized during construction, as well as the concrete reality of the resulting technoscientific system.[13]

Expanding beyond the specifics of his framework of analysis, Bijker argues that it is this very same constructivist conception regarding the malleability of technoscience that underlies any discussion of its politicization, while at the same time holding out the hope for a democratization of the process of scientific and technological decision making. "Without recognizing the interpretative flexibility of technology, one is bound to accept a technologically deterministic view," which conveys the image of autonomy, thereby denying the possibility of intervention. While this may seem obvious, at least within the constructivist framework, Bijker argues that it must be demonstrated in a rigorous manner, if we are "to escape the rather trivial level of observation that technology is man-made, and hence subject to many societal influences," for it is the very core of technology, and others would argue of science as well, that is socially constructed. At the same time the "obduracy" of technology must be recognized in terms of its ability to form "enduring practices, theories, and social institutions" with the very real resulting ability to then "determine" social development. The resulting "exemplary" sociotechnical ensembles, such as automobiles, for the many people who choose to adopt them, can subsequently take on what Langdon Winner has called "forms of life," which in turn tends to give rise to the idea of technological determinism. Nonetheless, the constructivist view holds out the reality of involving people in a democratic technoscientific decision-making process. Each group of citizens brings valuable experience to the process; it is not one limited solely to the technoscientific expertise of scientists and engineers. Because all technoscientific projects of any import involve sociopolitical aspects fully integrated with the technical, a wide variety of expertises needs to be involved very early on in the design process.[14]

Certainly there are those STS scholars who would disagree with the particulars of Bijker's notion of "technological ensembles" and those who would argue for either a more deterministic or a more relativistic interpretation of technoscience. Nonetheless, based on the preceding line of argument, I would argue, at least in the very general sense in which I have described it here, that the notion of "constructivism" is a core concept, perhaps even an authoritative body of theoretical work, that has come to characterize most of STS studies. In somewhat similar fashion, STS talks of the necessity to grapple with the "societal context" of both science and technology in order to fully understand, let alone act upon, them. Whether this moves STS in the direction of "interdisciplinarity" or "trans-disciplinarity" may not be of vital concern in many ways.[15] Certainly there are many barriers involved—for example, the separation of science studies from technology studies; the continuing disciplinary distinctions between philosophical, historical, sociological, and other approaches; and the frequent specialization of STS researchers versus the often more activist-orientation and generalization of STS education. Nonetheless, the concepts of constructivism and contextualism do seem to have provided a core intellectual understanding and tools of analysis that allow STS to serve, as anthropologist David Hess would have it, as "a site for public debates on issues of social importance."[16] It is in this location that a multiplicity of voices can come together, on the one hand to promote a better and more holistic understanding of science and technology as embedded in their societal context, and on the other hand to develop a more democratic process for technoscientific decision making. I will return to this latter theme at greater length in the final chapter, but next I want to turn to a brief discussion of where some of the key developments in STS seem to be taking place given just such a constructivist/contextual framework.

The Turn to Culture

In much the same way that the sociology of scientific knowledge approach took a "turn to technology," beginning in the late 1980s,[17] it would appear that science and technology studies is now making a "cultural" turn. Thus it is that David Hess, in his recent survey of STS studies scholarship, has pointed to the critical and cultural studies of science and technology as being the focus of the most current, interesting, and important research being done.[18] The very contextual nature of technoscience suggests the need not only to more fully understand the networks within which scientists and engineers pursue their work, but also to decipher the culture and values that give rise and meaning to them, and to analyze the differences in status and power accorded to some actors, theories, and systems while denying the same to others. Among the more fruitful frameworks in this regard are those of Cultural and Critical Studies.

Cultural Studies is an interdisciplinary inquiry that draws upon theoretical frameworks such as Antonio Gramsci's notion of "hegemony," focuses on popular culture and subcultures, and utilizes a range methodologies including fieldwork, ethnographic study, and archival research. It also tends to concern itself with feminist, antiracist, and postcolonial issues in which issues of power and status are foregrounded and as such it is politically engaged. Critical Studies of science and technology somewhat overlaps Cultural Studies by drawing on feminist and antiracist studies, but is also draws more overtly on the radical science movements of the 1930s to the 1940s and the 1960s such as the British Society for Social Responsibility in Science and the literature critical of modern technosociety (see chapter 1). Together Cultural and Critical Studies thus concern themselves with issues of democracy and social justice.

STS scholars frequently utilize the concept of "hegemony" associated with the work of Italian Marxist Antonio Gramsci to explain the distribution of power and wealth even within democratic societies. In this view the ruling class establishes a system of values and beliefs generally accepted by and institutionalized within society, yet it also allows for other value and belief systems to coexist.[19] This provides an illusion of democracy, which is not threatening to support for the hegemonic system as long as the detractors are marginalized. In this regard, traditional rational science can be viewed as a hegemonic force as it suppresses alternative viewpoints, often characterizing them as pseudoscientific or antirational. Hess suggests the way government and large corporations can marshal well funded, mainstream science on their behalf in the face of environmental critics as exemplifying this tendency. The way government and big business have tended to constrain the environmental movement's concerns regarding global climate change could be seen as just such a specific example. Andrew Webster makes a somewhat similar point, when he notes how "elite decision-makers" often set political agendas "*in advance* of debate" in order to set "manageable" limits for the terms of discussion "within the context of the status quo."[20]

Another concept that Hess suggests has been widely adopted by critical analysts of science and technology is that of "reification" drawn from the work of Marxist scholar Georg Lukas. In this view, social and human relationships are transformed, or reified, into commodities and things; thus, labor is no longer creative, but rather merely a commodity to be sold, although hopefully at least to the highest bidder. Hess points to the work of antiracist and feminist scholars, such as Donna Haraway, who show how science can reify certain values and categories, such as connecting or attributing intelligence or rationality to differences in race or sex, making them appear "natural," when indeed that is not the case at all.[21]

Feminist scholars have not only extended existing critical frameworks of science and technology analysis, but have also contributed some of their own, most particularly with the concept of "gender" developed originally to counter arguments that social action could be reduced to differences in biological sex. Donna Haraway thus describes gender as "a concept developed to contest the naturalization of sexual difference." More recently, some feminist scholars, such as Ruth Hubbard and Marianne Van den Wijngaard, have argued that even some previously unchallenged biological understandings of sexual differences are, in fact, culturally gendered. Van den Wijngaard's work "reveals how biomedical scientists reinvented the sexes and how they assigned new and different meanings to gender, masculinity, and femininity in their investigation of sex hormones"; sex thus turns out "to be more culturally determined than scientists had initially imagined." She concludes that "absolute masculinity and femininity do not exist but that these categories are constructed truths about bodies and psychological characteristics ascribed to females and males." Hubbard, trained originally as a biologist, argues that "we live in dynamic interaction with our environment," and thus that "our biological and social attributes are related dialectically." For her, "Sex differences are socially constructed because being raised as a girl or a boy produces biological as well as social differences." It is in this regard that "women's biology is a social construct and political concept, not a scientific one." Dianne Hales, an M.D., takes a somewhat broader view in that she does not deny women their distinctive biology, while at the same time recognizing life's constructed nature. She argues that, "Gender is only a part of our identity, and each woman's life is shaped by myriad factors—among them genetic endowment, innate temper, childhood experience, physical health, race, culture, ethnicity, sexual orientation, and economic status."[22]

Among a number of other concepts utilized in feminist science and technology studies has been that of "essentialism"—the idea that there are essential distinctions between men and women that can be attributed to and associated with biological differences between the sexes. While some feminists such as Susan Griffin and Hillary Rose have argued that because women bear responsibility for childbearing and rearing, they are somehow closer to nature, and hence would create a more humane, subjective, and "empathetic" science, sensitive especially to women's values,[23] recent scholarship has criticized this viewpoint. Drawing on scholarly evidence such as that offered by anthropologist Marilyn Strathern, they argue that not only are science and technology socially constructed, but so too are gender values themselves. Philosopher of science Sandra Harding, while arguing that gender is crucially important, notes that the effect of historical research is "to challenge the universality of the particular dichotomized set of social behaviors and meanings associated with masculinity and femininity in Western culture." She suggests instead a variety of feminist perspectives, including those ways that women's experiences are affected by race, class, and culture.[24] Scholars such as sociologist Judith Wajcman have extended the "antiessentialist" position beyond science to the realm of technology studies by examining such issues as household and reproductive technologies. Interestingly, although not surprisingly so, given current feminist thinking regarding the diversity of individual experiences with technology, Wajcman and others have suggested that such technologies, but especially those associated with reproduction, are at once liberating and empowering as well as alienating and disempowering.[25] Thus, most feminists now view that any potential substitution of a "female" science or technology for the existing male-dominated science would but substitute one set of exclusionary values and practices for another, when what is called for is a collective, but nonoppressive vision that recognizes the constructed nature of technoscience, including its gendered context.

Antiracist studies of science and technology have also contributed important critical insights during the past decade. Once again, Donna Haraway's study of primatology shows how, even within the coalition of "oppositional" women primatologists who ultimately brought new understandings and a restructuring of the field, there were hegemonic distinctions, especially of white women over women of color. The issue of environmental racism has also been highlighted, especially in the work of Robert Bullard, most notably in his book *Dumping on Dixie*. In this study of the location of hazardous waste facilities, Bullard shows how the hegemonic powers of business and the state have knowingly and willingly sited such facilities near minority and economically disadvantaged communities at disproportionate levels given the population of such groups. That such decisions are often made with the concurrence of the communities affected, for example, due to the desperate need for jobs or the difficulties of assessing the scientific and technical issues involved, does nothing to undermine the hegemonic argument regarding issues of racial, social, and economic justice. Rather, they only point to the difficulties encountered in unpacking and assessing the dynamics involved in such issues.[26]

One area that has recently begun to receive more attention is that regarding the ways poor and developing nations deal with the development and transfer of modern, especially Western, technoscience. David Hess in *Science and Technology in a Multicultural World* takes as a starting point "the culture concept" and asks "the question of how to include historically excluded perspectives" as an attempt to move beyond a theoretical social construction of scientific knowledge and technology to "adopt a culture-and-power perspective." In somewhat similar vein, Sandra Harding is interested in epistemological issues associated with science and technology in today's multicultural and postcolonial world. In her book *Is Science Multicultural?* she counters the "singularity" of traditional internalist epistemological accounts of science, with a view of all science projects as culturally situated, "local knowledge systems." That is, in contrast to a view that posits the existence of but one nature with an accompanying truth about it, she is arguing that modern European sciences and technologies are not the only accomplished ones, nor are they "universally" applicable. She believes, instead, that there can be a "continuum" of useful standards of knowledge ranging from "global" to "local" depending on cultural context. It is important to note here, especially in light of some issues to be discussed below, that Harding is not proclaiming an epistemological relativism. She is seeking instead what she calls a "strong objectivity program"—the notion that "different cultures' knowledge systems have different resources and limitations for producing knowledge; they are not all 'equal,' but there is no single possible perfect one, either." Ultimately it is her hope "to improve the lot of the majority of the world's peoples" in terms of global democratization, with "a more adequate . . . understanding of how sciences do and could produce knowledge."[27]

In addition to the insights gained from feminist, antiracist, and postcolonial studies, anthropology has increasingly come to offer important contributions that complement those coming out of history, sociology, and philosophy. One of the strongest proponents of the value of anthropological perspectives to current cultural and critical studies of science and technology has been David Hess. In explicating the contributions of anthropology to the interdisciplinary STS mix, he clarifies the adoption of several key concepts, each of which is tied to the recent turn toward cultural analysis. First is the use of the term "ethnography," which he suggests has been used loosely in science studies to refer to fieldwork-based methodology, when in anthropology the term implies much more. In the latter usage, the ethnographer would expect to spend longer periods of time learning appropriate language skills, developing contacts with informants, and conducting participant-observation in the community or unit being studied. Thus, Hess suggests that such early studies as Karen Knorr-Cetina's 1981 *Manufacture of Knowledge* and Bruno Latour and Steve Woolgar's 1979 *Laboratory Life,* which were more limited in focus and time frame, and tended to utilize sociological methodology, would be more properly called "laboratory studies," as, in fact, they often are. In contrast, the "second wave" of ethnographic studies "works with larger field sites such as transnational disciplines or geographic regions, addresses questions defined largely by a concern with various social problems (e.g., sexism, racism, colonialism, class conflict, ecology) that are theorized with hybrid feminist/cultural/social theories," and involves those with anthropological training to a much greater extent. Hess also points to the differing usages of the term "culture" versus that of "society." For most sociologists, historians, and political scientists, the latter would be the more all-encompassing term referring to the state or community, its economy and social structure, with culture referring to the values and norms responsible for socialization and self-maintenance of a given unit. For anthropologists the reverse is the case, referring to "the total learned knowledge, belief, and practices, both conscious and unconscious, of a social unit." Thus, "culture permeates and includes all social institutions and practices," and it is "contested, changing, and distributed." As such there may be contrasting subgroups within a given culture, each with differing values and expertises. Finally, Hess notes that current anthropologically based work tends to focus on larger units or "arenas" of analysis, than the earlier, network-focused laboratory studies. Such arenas might thus include not only the scientific and technical areas, but also the political, corporate, and social realms as well.[28]

A good example of the new direction in which science and technology studies research is headed is the ethnographic-oriented body of work on the physics community that Sharon Traweek has published. In her first book, *Beamtimes and Lifetimes*, Traweek compared Japanese and U.S.

physicists, looking at such issues as national cultures, gender roles, social groupings and career patterns, and equipment design and usage. This and subsequent work has allowed her to understand how the influence of cultural differences can lead to different research questions and methodologies, education and training, and funding patterns, that in turn can lead to contrasting laboratory equipment designs.[29]

Another area of recent, culturally oriented STS research that bears mentioning is that dealing with broader public understandings of science and technology (PUST). Traditionally the public has been perceived as being the problem in terms of understanding science and engineering, which stems in large part from a view that privileges technoscientific training and expertise and, in turn, leads to calls for enhanced scientific and technological "literacy" on the part of a "problematized" public.[30] Thus, it should not come as much of a surprise that two of the three main areas of research into PUST, as identified by Brian Wynne in his summary in the 4S *Handbook,* tend to accept fairly traditional positivist images of science and scientific knowledge and problematize the public. Thus, large-scale quantitative surveys of public scientific understanding, such as those included in the U.S. National Science Foundation's "science indicators" program, generally fail to deal with the ways the public does "understand" the nature of scientific process, institutional characteristics, and social implications, even if people do not fully comprehend specific scientific content. Similarly, analysis of the "mental models" by which lay people organize scientific knowledge has often tended to accept without much question the standard norms of what constitutes science. In contrast, constructivist-oriented research, with methods derived from anthropological-based ethnography, participant observation, and in-depth interviews, "examine[s] the influence of social contexts and social relations upon people's renegotiation of the 'science' handed down from formal institutions. . . . " In this way the assessment of what counts as technoscientific knowledge, or the manner in which a technoscientific issue is defined, are themselves problematized, thereby, in turn, opening up the technoscientific process to much greater public understanding and participation, an issue I will return to in the final chapter.[31]

Wynne suggests that the public always encounters science and technology in some sort of social context, as a result of which they may ignore that science which they deem irrelevant, or may fail or be unable to act on available scientific knowledge, even if they do not dispute scientific experts. Lack of trust regarding the relevant institutions, not just the technical information available, is also an issue of importance. None of this means that members of the public are necessarily scientifically illiterate. Approaches to policy issues entailing public understanding that take into account local context and that involve a willingness on the part of technoscientific institutions to examine reflexively their own assumptions and commitments can assist in negotiating what counts as "good" science and technology and lead to better resolution of the issues at hand. Two recent STS studies that are at once revealing of current research in this regard and suggestive of the practical applications of such research can be found in Brian Wynne's own study of local understandings of radiation pollution originating from Chernobyl as it affected sheep farmers in the Lake District of England and in Steven Epstein's study of how the lay AIDS community contributed to the development of alternative medical treatments and procedures.[32]

In the first case, it was discovered that high levels of radiation were present in the sheep in the Cumbrian fells area of England, following heavy rains that coincided with the cloud of debris that had circled the globe following the April 1986 Chernobyl nuclear explosion. Despite initial government pronouncements that there were no health hazards, it quickly became evident that radioactive material had been taken up in the grasses upon which the sheep fed, in turn contaminating the lamb meat. Government officials quickly initiated what was expected to be but a short ban on the movement and slaughtering of sheep, based on assumptions that no further radiation would be taken up by the grasses. Unfortunately this proved mistaken due to differing soil types, some of which allowed radioactive caesium to be continually absorbed. The result was an extended ban, which led to a direct threat to farmers' incomes because they could not sell their annual crop of lambs, except at a loss, and an indirect concern for overgrazing limited pasture land if the lambs could not be slaughtered or moved. Scientists, convinced of their own expertise, failed both to recognize the complexities of hill area sheep farming and to credit local knowledge regarding ecology, soil type, water accumulation, and grazing patterns. In the minds of local farmers, especially in the face of severe financial loss, it should have been possible to develop appropriate restrictions that drew upon scientific "expertise," while taking into account the need for flexible management based on lay knowledge, which was substantial.

In the second case, the AIDS community, especially in San Francisco, was becoming increasingly frustrated at "official" policies that denied patients access to alternative treatments which, although not fully tested, showed promising results; by medical research protocols that denied research subjects access to their normal medications during the trial; and by testing procedures that required base groups who received placebos, in effect preventing them from receiving possibly beneficial drugs. In the mid-1980s, the potentially promising AIDS drug AZT (azidothymidine) was undergoing controlled placebo tests. Activist AIDS research groups and publications began pressing the medical establishment to advance timetables and utilize procedures other than placebo trials that, in effect, led to some people dying. In short order the AZT test proved so effective that the initial study trial was suspended and the drug made available to patients. As scientific understanding increased among AIDS patients and activists, they increasingly began to work with community doctors to develop alternative trials of drugs, such as aerosolized pentamidine for the treatment of PCP pneumonia. Such trials among community-based groups in San Francisco and New York accelerated the timetable and eliminated the practice of using placebos. Ultimately the data collected was deemed so reliable that the FDA approved the use of the drug. Over the next several years AIDS activists increasingly sought to work with scientists, whom they believed did not fully appreciate the perspective, and knowledge, of those actually affected by the disease. Among the concerns voiced were the medical establishment's over-focus on AZT, which was highly toxic and not really a cure; increased access to new drugs outside normal channels; and the necessity of changing clinical trials to make them "more humane, relevant and more capable of generating trustworthy conclusions."[33] As activists gained increasing knowledge regarding AIDS and the associated science, physicians increasingly came to respect their ability to coparticipate in the research process, especially with regard to establishing more reliable clinical trials.

In both cases, the STS studies have shown that, in fact, members of the lay public acted very intelligently. By learning the necessary underlying science, as well as questioning the hierarchical status of received scientific wisdom, the AIDS community was able to work constructively with the established scientific community to effect changes that all involved agreed were improvements. In contrast, there was less open give and take between British scientists and local sheep farmers. If the former had heeded the insights of local lay knowledge, they might well have avoided at least some of the frustrations engendered there. These studies reveal then that, at least with regard to issues that concern them intimately, the public is capable of developing a sophisticated understanding of the technoscientific knowledge, as well as the societal context, involved. They also point to the wisdom of involving the public in technoscientific decision making at as early a stage in the process as possible.

In summary, it should be clear, even from this limited number of examples, that the current focus of studies of science and technology shows how "general cultural values," such as those associated with nationality, class, gender, race, and ethnicity help to shape technoscience beyond the earlier constructivist insights regarding the way "particular" values influence research problem selection or theory preference. As David Hess nicely puts it, "The question is not how to eliminate cultural values but instead how to find out which cultural values structure science and whether different or better sciences would result if other values replaced the ones currently in place."[34] In that search I believe STS has a very real role to play, a theme I will return to in the final chapter.

Science Wars

One unfortunate result of all this intellectual theorizing about the socially constructed nature of technoscience has been the strong backlash from certain quarters of the scientific community regarding the accuracy and appropriateness of at least some of the claims being made by STS scholars. I say unfortunate, not because it is inappropriate or unimportant to confront directly and reflexively one's own analysis of the subject matter, but rather because of the polemical sharpness and at times ad hominem tone of much of the debate, when what would be more fruitful would be a more reasoned discussion of the central issues at hand. While much of this criticism would appear to miss, or ignore, the central focus and insights of STS, certainly much toner and intellectual blood has been spilled of late, and therefore it requires at least brief mention, if this discussion of the current state of STS scholarship is to be complete. Thus, let me turn now to what has become known as the "science wars."[35]

Many scientists hold tightly to the traditional ideal of objective knowledge based on reason and empirical evidence. For such individuals, relativist claims that scientific knowledge is socially constructed and, in the most extreme interpretations, not to be found in an objective autonomous nature, but rather as the result of a set of historically and culturally elaborated set of conventions, is profoundly unsettling, if not threatening. Faced with very real research funding cuts, such as that represented by the 1993 congressional decision regarding the Superconducting Supercollider, apparent widespread scientific illiteracy among schoolchildren, and widely held pseudoscientific beliefs on the part of the general public, some scientists have viewed much of STS as antiscience, "leftist," and part of a much broader, academically inspired, postmodernist decline into cultural decay. In defense of the objective nature of scientific knowledge and science as a special way of knowing, a number of such individuals have lashed out with sharp attacks on constructivist critiques of science.

The first major volley in this so-called science war was that of Paul Gross and Norman Levitt in their 1994 book *Higher Superstition: The Academic Left and Its Quarrels with Science*.[36] Gross, a biologist, and Levitt, a mathematician, loosely categorize those humanists and social scientists whom they see as "misreading" and "disliking" science as the postmodern "academic left" and accuse them of "intellectual dereliction." For Gross and Levitt there can be no doubt that "science *works*" and anyone who challenges that notion is "antiscience." Seemingly, they brook little questioning of science and hence look harshly on "radical" feminist, racial, environmental, and relativist critiques, although they do recognize a "weak" constructivism. Thus, while they are willing to accept science as a "social construct" in terms of how "interests, beliefs, and even . . . prejudices" can influence research problem selection, funding, and recognition, they draw the line when it comes to epistemological skepticism. They hold fast to the view of science as "a body of knowledge and testable conjecture concerning the 'real' world." In their view science is "a reality-driven enterprise." The problem appears to be that despite a disclaimer that this is "a book about politics and its curious offspring, not about epistemology," Gross and Levitt generally tend, in fact, to limit their focus to epistemological sorts of issues rather than to the social dynamics and political implications of science.[37] As a result, they tend to indiscriminately brand all those who raise questions about the social efficacy of science, as well as those extreme relativists who may, in fact, challenge the objectivity of science, as being antithetical to what they see as the progressive liberating nature of science. Most STS scholars are not "antiscience" at all, but seek, as do Gross and Levitt themselves, to try to understand the scientific endeavor and to extend democratic participation in the public debate surrounding technoscientific issues.

One difference would appear to be what counts for scientific "literacy," and here Gross and Levitt strike out at those who they believe fail to understand science "at a deep level," a task requiring "much time and effort." Such effort they view as being incommensurate with "the style of education and training that nurtures the average humanist." Thus, they take to task such scholars as cultural critic Stanley Aronowitz, sociologist Bruno Latour, historians Steven Shapin and Simon Shaffer, and feminist theorists Sandra Harding and Donna Haraway. By dissecting selected quotations or limited aspects of specific works that many STS scholars would argue were either admittedly extreme or unrepresentative of the field, Gross and Levitt generalize about the field as a whole. Thus, they are quick to highlight Latour's "Rule of Method" that states "Since the settlement of a controversy is the Cause of Nature's representation, not the consequence, we can never use the outcome—Nature—to explain how and why controversy has been settled," but fail, as Brian Martin points out in his review of their book, to note that many STS scholars themselves have critiqued Latour's stance, or that Latour himself seems to have more recently backed away from the position. Similarly, they critique but two sentences from Sandra Harding's *The Science Question in Feminism*, viewing them as characteristically hostile and representative of the feminist critique as a whole. Independent of whether one agrees with these two assessments or other examples, and certainly some STS scholars do, what Gross and Levitt unfortunately do is tend to castigate the field as a whole, characterizing it as mere academic "higher superstition." They thereby negate many positive constructivist insights and ignore many of the contributions of the science policy and activist components of STS.[38]

Not surprisingly perhaps, Gross and Levitt's book opened the floodgates for further attacks and counterattacks. Joined by geographer Martin Lewis, Gross and Levitt themselves assembled a group of some forty speakers at a 1995 New York Academy of Sciences–sponsored conference to further warn about the dangers in abandoning the ideal of scientific objectivity. The resulting proceedings volume, *The Flight from Science and Reason*, includes positions ranging from Henry Greenberg's introductory gambit in which he associates the recent growth of HMOs (health maintenance organizations) with the "social construction of reality [which] has come to medicine," to critiques of the excesses of "radical" feminists and environmentalists by philosopher Janet Radcliffe Richards and geographer Martin Lewis respectively, to philosopher Mario Bunge's McCarthy-like suggestion that "constructivist-relativists" are but a "'postmodern' Trojan horse" within the academic citadel, which leads him to call for a "Truth Squad" to "expel the charlatans from the university."[39] There are constructive contributions such as that of biologist Meera Nanda who, somewhat ironically, draws in part upon Donna Haraway's idea of "situated knowledge." In this view, local traditions of knowledge and "emancipatory" Western scientific rationality can "exist in creative tension with each other," interacting to provide "partial but reliable and verifiable knowledge" supportive of the emancipation of postcolonial people.[40] This is somewhat ironic in the sense that in their book *Higher Superstition*, Gross and Levitt include Haraway in what they call the "feminist-critique-of-science mafia."[41] In the end, it is again, as with *Higher Superstition*, not so much a question of there being no useful perspectives or insights offered in these essays, but rather that the overall tone of the volume blackwashes STS as though there were no variation within the field and no value whatsoever to be found in constructivist viewpoints. This is as much a "flight from reason"-able discourse as any that can be found in STS.

At roughly the same time Gross, Levitt, and Lewis were assembling their "Flight from Reason" conference speakers, Andrew Ross and Stanley Aronowitz, coeditor and cofounder, respectively, of the cultural studies journal *Social Text,* were compiling a set of science studies essays in direct response to the "churlish tone" of *Higher Superstition*'s "choleric attack."[42] The volume includes a wide range of authors, a number of whom, such as Sandra Harding and Aronowitz and Ross themselves, bore the brunt of Gross and Levitt's frontal assault. One of the common themes that runs throughout these essays is that STS has a much wider and more varied set of concerns than the supposed "antiscience" image in which it is portrayed by Gross and Levitt. Ross suggests the varied aims include: a desire "to provide an accurate scientific description of empirical scientific practice," a wish "to see science redeem its tarnished ideals from internal abuse and external impurities," an attempt to "persuade scientists to be self-critical about the political nature and social origins of their research and to engage in advocacy science to combat the risks and injustices that are side effects of technoscientific development," and a call "to create new scientific methods that are rooted in the social needs of communities and accountable to social interests other than those of managerial elites in business, government, and the military."[43]

Langdon Winner similarly sees four different "projects" as inspiring STS thinking and research. One seeks "simply to understand how modern science and technology work, how their various practices, institutions, and tangible products have developed." A second involves the application of traditional social science and humanities disciplines to the study of science and technology, topics traditionally seen not to be "amenable" to such methodologies. Third is "the need to respond to a host of practical problems" engendered by changing scientific knowledge and technological developments. Lastly, Winner notes the critical concerns of philosophers and social theorists regarding the "profound crisis in the underlying conditions of modern life and thought" to which science and technology have contributed.[44]

In addition to indicating how STS is not a monolithic conspiracy, most of the contributors are also quick to point out that STS is not the "antiscience brigade" that Gross and Levitt would suggest, nor is it a bastion of extreme relativism. This is not to say, however, that STS, at least in some of its guises, does not question the authority of science as a special privileged way of knowing, does not reveal the tendencies toward commercialization and cultural domination in modern technoscience, and does not show how politics and the role of power infuse science and technology. Sandra Harding argues not that "the sciences are epistemologically relative to each other and every culture's beliefs . . . equally defensible as true," but that "they are historically relative to different cultures' *projects*." Thus, "If you want to do modern agribusiness, modern technosciences can help; if you want to maintain a fragile environment and biodiversity, those sciences, so far, have been of little assistance."[45] It would appear that it is such observations and interpretations as these that, in fact, pique the ire of Gross, Levitt et al., not the ostensible antirealist and antiscientific stance they accord to STS scholars.

One of the more bizarre, but at the same time revealing, events to evolve out of the science wars has been what has become known as the "Sokal hoax." Alan D. Sokal, a physicist at New York University, had submitted a manuscript entitled "Transgressing the Boundaries: Toward a Transformative Hermeneutics of Quantum Gravity" to the non–peer reviewed, cultural studies journal *Social Text*. In itself, especially given the uproar that was to follow, this was an interesting choice of journal to consider for submission, for most STS scholars likely would not see it as one central to the field. Nonetheless, after some apparent hesitancy and internal debate, the editors, Andrew Ross and at that time Bruce Robbins, decided to accept the unsolicited article, extremely ironically given what would follow, for their special issue on the "science wars" compiled in response to Gross and Levitt's *Higher Superstition*. The irony is twofold. On the one hand it stems from the fact that, as Sokal readily admitted in a essay published in *Lingua Franca* at approximately the same time, the article was a complete parody, full of satirical, postmodern "gibberish," yet it was ultimately published in a special issue designed as a response to *Higher Superstition*. On the other hand, this very same issue includes numerous articles reflecting the broad range of thinking that goes on under the STS rubric, of which Sokal, Gross, and Levitt seem to be unaware or choose to ignore. Sokal had been motivated by what he viewed as "the proliferation, not just of nonsense and sloppy thinking, but of a particular kind of nonsense and sloppy thinking: one that denies the existence of objective realities." Like Gross and Levitt before him, Sokal was concerned that "losing contact with the real world . . . undermine[s] the prospect for progressive social critique." Like Gross and Levitt, Sokal believes in the progressive natures of science and technology and derides what he sees as a relativist-inspired postmodern decline.[46]

The Sokal parody has been received variously depending on one's position regarding the broader "culture wars" of which this skirmish is but a small part. For those in the scientific community convinced of the irrationality and muddleheadedness of science studies scholarship, it was proof positive of "the fraudulence of certain leading figures in cultural studies."[47] Some people decried the deception that Sokal utilized to promote his hoax, while for many in the science studies community, Sokal's piece displayed his failure to fully grasp the complexities of the constructivist literature. Yet others criticized the editors of *Social Text* for sloppy, if not irresponsible, editorial review, which undermined the integrity of STS as a field of study.[48]

Flush with "success" from this venture into science studies criticism, Sokal and another physicist colleague, Jean Bricmont, who teaches at the University of Louvain, extended the critique in a book-length version, entitled in its English-language translation, *Fashionable Nonsense: Postmodern Intellectuals' Abuse of Science*. Published originally in French as *Impostures Intellectuelles*, the book takes on what Sokal and Bricmont view as the "abuses" against science, the "fashionable nonsense," perpetrated primarily by French intellectuals such as psychoanalyst Jacques Lacan, sociologist/philosopher Jean Baudrillard, and sociologist of science Bruno Latour. The authors view these French intellectuals as also having influenced American science and cultural studies. Thus, much of the book is given over to analysis of passages about science taken from the works of these authors, with an eye toward rectifying their misconceptions and misapplications. They also target "epistemic relativism," the view that "modern science is nothing more than a 'myth' . . . or a 'social construction.'"[49]

Sokal and Bricmont readily accept the idea that scientific activity takes place within a societal context, a "weak" sort of constructivism to use Gross and Levitt's term, but they wish to draw the line sharply in front of those who argue that only social causes and not Nature itself can be used to explain scientific "content." Thus, while they recognize we can only approach the world through our senses, that "we have no *proof*," that anything exists outside those sensations, "it is simply a perfectly reasonable hypothesis." In their minds, "The best way to account for the coherence of our experience is to suppose that the outside world corresponds, at least approximately, to the image of it provided by our senses." While science cannot be "codified in a definitive way, it is "a rational enterprise." Thus it is that Sokal and Bricmont can agree with philosopher Paul Feyerabend when he argued in his book *Against Method* that science cannot be "run according to fixed and universal rules," yet heartily reject as "erroneous" his subsequent statement that, "All methodologies have their limitations and the 'rule' that survives is 'anything goes.'" Among other relativist arguments, they also critique Bruno Latour's Third Rule of Method from his *Science in Action* (see above), where he argues that Nature cannot be used to resolve scientific controversies. For Sokal and Bricmont, it is "not enough to study the alliances or power relationships between scientists," but what is also needed is "a detailed understanding of the scientific theories and experiments," one that does not confuse the existence of "facts with *assertions* of facts."[50]

What is interesting and perhaps surprising given the tone of much of this debate, although it should be noted that Sokal and Bricmont are somewhat less polemical than Gross and Levitt, is that at root none of them are that far apart from the position of most STS scholars on the constructed nature of scientific knowledge, the existence of an obdurate reality, and the very real problems society confronts in its development and use of technoscience. Most STS scholars are not "radical" constructivists who simplistically deny all "reality." This is why no one took up Sokal's satirical invitation in his *Lingua Franca* revelation to step out of his twenty-first-floor apartment window, if they really believed "the laws of physics are mere social conventions."[51] Rather, most STS people are "moderate constructivists" or "amiable constructivist realists" as David

Hess and Bruce Lewenstein have put it.[52] That is, most believe scientific *knowledge* comes both from nature and from social forces and cultural variables or, in other words, that the reality of Nature constrains what we can know, but that *knowledge* is integrally shaped by sociocultural influences. Given this seeming core of agreement over moderate social constructivism and a desire on all sides to improve at once our understanding of technoscientific processes and the decision making that surrounds them at all levels, much of the venom directed at sciences studies scholarship would appear misdirected in several ways. It is misdirected in that many of the Gross-Levitt and Sokal-Bricmont targets are not people who would generally be viewed as a part of STS, and for those more extreme relativists who are, they are but a minority. At the same time, some of the more important problems related to technoscience recognized by both pairs of authors—militarism, sexism, environmental impacts[53]—largely go untreated in their work, except with regard to caricaturing and dismissing out of hand, without seeing any value in them whatsoever, certain admittedly radical stances.

For these reasons, there is perhaps good reason not to view the "science wars" as being of central concern within the STS movement as a whole, or even within the more limited academic wing of STS scholarship, even though they have occupied the attention of a good many people since the mid-1990s. In fact, there may be some evidence that the combatants are beginning to talk to each other more reflexively and constructively, including for example a 1998 "debate" between Sokal and Latour.[54] Nonetheless, there continues to be a stream of books and articles attempting to grapple with the issues at hand. Thus, Timothy Lenoir edited a volume entitled *Inscribing Science: Scientific Texts and the Materiality of Communication*, which draws on the inspiration of postmodernist theorists like Jacques Derrida as a direct response to the Sokal affair, while on the other "side," Noretta Koertge has edited a volume entitled *A House Built on Sand: Exposing Postmodernist Myths about Science*, which contains among its essays one by Sokal—"What the *Social Text* Affair Does and Does Not Prove."[55]

Independent of whether these disputes are fading or thriving, it seems important that neither the scientists nor the STS scholars and activists wall themselves off. Historian of science and technology Bill Leslie, in a plenary address at the 1999 annual NASTS meeting entitled "Who's Talking? Who's Listening? Who Cares? Reestablishing a Conversation in STS," suggested that it was the "Reestablishing a Conversation" part of the title that was most important. In his view STS is at its best as a "dialog," where a "conversation" among STS scholars, scientists, students, and the general public can be conducted.[56] If STS finds itself overly focused on issues of scientific epistemology, it runs the risk of not dealing adequately with the larger societal and political implications of technoscience. It is also important, however, to recognize that one important question underlying the otherwise largely STS academic exchanges has to do with who gets to say what about technoscience, an issue to which I will turn my attention in this chapter with some brief concluding thoughts and then elaborate upon further in chapter 5.

Conclusion

One of the main concerns of Gross and Levitt and of Sokal and Bricmont is that analysts of science and technology have an adequate if not a deep understanding of what they study. In this regard, they are intensely bothered by perceived attacks on rationality as evidenced by religious and New Age "superstitions"—creationism, telepathy, astrology—as well as what they refer to as "ecobabble," "radical feminism," "Afrocentrism," and "alternative modes of healing." For Gross and Levitt and for Sokal and Bricmont, this is nothing but an "amalgam of quackery and self-delusion," amounting to "sheer puffery." In their minds such superstitions and causes entail a "flight from reason" and represent nothing but "fashionable nonsense." At the same time they are also bothered by what they see as the intellectual "dereliction," "posturing," and "dishonesty" of those "radical" scholars of the "academic left" who, in their view, have the temerity to critique science, "*without* troubling to become well informed about the substance and the inner logic of the scientific enterprise." In their minds, active citizenship in technoscientific affairs "requires a usable knowledge of science and technology." Beyond this, they view it as certainly legitimate to study philosophically the content of the natural sciences, but to do so, "one has to understand the relevant scientific theories at a rather deep and inevitably technical level." Such prerequisites are not at all unreasonable, and I would argue precisely what most STS scholars attempt to do, and especially so when analyzing the content of science.[57]

Given that Gross and Levitt and Sokal and Bricmont recognize the societally problematic nature of modern technoscience, it is unfortunate that they seem to confuse legitimate sociopolitical concern for such with perceived assaults on "objectivity," which they then term "antiscience." In point of fact, few STS scholars, if any, are "antiscience," or "antitechnology" for that matter, including even those who are primarily concerned about the negative impacts of the latter upon society. What they do attempt to do is to better understand the natures of science and technology and how scientists and engineers create and apply their bodies of knowledge. They do this as an intellectual end in and of itself, but also for many with an eye toward enhanced technoscientific decision making. Thus, theirs is not an attack on "objectivity" but rather the application of a constructivist view that sees technoscience as societally shaped, albeit within the constraints of a material reality, to create the technological "ensembles" of which Weibe Bijker speaks, to refer back to but one such scholar already mentioned.

Sandra Harding, whose work often serves as something of a lightning rod for the anti-STS critics, is herself careful to claim in the introduction to her edited book, *The "Racial" Economy of Science,* that she and the contributors are not engaged in "science bashing" nor is the collection itself "against scientists." Rather, the authors "value scientific knowledge and the resources it can provide when harnessed to projects of making and re-making democracy."[58] What Harding wants to argue is not that there is no "reality," but that scientists always understand "nature-as-an-object-of-knowledge." In this sense, there can be no "pure" nature, or for that matter "pure science," for scientists always bring to their observations traditions, values, and socially constituted theories. As she notes in *Is Science Multicultural?* "Of course 'there is a world out there.'" For Harding certainly "reality exists"—the law of gravity affects us all. What is at issue, however, is whether there is "one and only one best way" for science to represent "nature's order." She believes that, although they may bear "a relation to reality," "no single map or collection of them can perfectly reflect 'reality.'" Instead, different standards of knowledge depending upon local circumstance and project can help to "maximize objectivity." In what she calls a "strong objectivity" program of analysis, Harding expects, by drawing on "external" standpoints or the perspectives of various marginalized groups, to be able to "improve the performance of the sciences, natural and social," by generating accounts of nature that are "less false"—and more democratic. This "does not mean giving up the epistemology of modern science completely," although it does require it to be "updated." She much prefers a notion of "co-construction," in which technoscientific ideas and principles, or "projects" in

Harding's terminology, "co-evolve" with "elements of their particular social formation," the whole thing "'constrained' by nature's order. . . ."[59]

Given that the critics of STS appear to accept at least something approaching just such a moderate constructivism themselves, one is left wondering what all the fuss is about. At least some of it would appear to stem from the STS movement's "problematization" of science and especially technology, which undercuts traditional views of science as value neutral, and hence may appear challenging, but this in itself is neither nonrational, nor antiscience. Following this, part of the confusion, and resulting tension, would appear to come from the tendency on the part of critics of STS to misinterpret the wide range of STS analysis as though the only issue was one of epistemological content, when in fact what is at stake and under analysis is the much broader social dynamics of science, and of technology.

To resolve the technoscientific problems and issues in such areas as the environment, health, and today's postindustrial global economy will require a working together to understand the reciprocal relationships among science, technology, and society. It will require cooperation and the shared insights of a coming together of scientists, engineers, and businessmen, as well as historians, philosophers, sociologists, and anthropologists. And it will benefit from integrating the insights of feminists, antiracists, and postcolonialists. In short, it will require a multidisciplinary, if not interdisciplinary, approach to understanding the societal context of science and technology at both the intellectual and activist levels. To achieve that end will also necessitate problematizing science and technology in the sense of exposing their value-laden natures, not just problematizing the public's level of technoscientific literacy, important as that task may in itself be. This, however, is not the same as completely denying the rationality of science or adopting a radical relativism. Thus, for critics of STS to polemically assault the field as "antiscience" confuses issues and hence misses the mark, and widely so. Such misdirection, in fact, only repolarizes C. P. Snow's "two cultures," when what is called for instead is further interdisciplinary dialog.[60]

Notes

1. Julie Thompson Klein, *Interdisciplinarity: History, Theory, and Practice* (Detroit: Wayne State University Press, 1990).

2. Gary Bowden, "Coming of Age in STS," in *Handbook of Science and Technology Studies*, ed. Sheila Jasanoff et al. (Thousand Oaks, Calif.: Sage, 1995), 64–79.

3. Bowden, "Coming of Age in STS," quotations, 71–72, 75, 64. It is interesting to note here that Leo Marx in a review of Stephen H. Cutcliffe and Robert C. Post, eds., *In Context: History and the History of Technology—Essays in Honor of Melvin Kranzberg* (Bethlehem, Pa.: Lehigh University Press, 1989) raised a similar question whether, given the "triumph" of contextualism regarding technology, there was a "rationale for making 'technology,' with its unusually obscure boundaries, the focus of a discrete field of specialized historical scholarship." See *Technology and Culture* 32 (1991): 394–96.

4. Henry H. Bauer, "Barriers against Interdisciplinarity: Implications for Studies of Science, Technology, and Society (STS)," *Science, Technology, & Human Values* 15 (Winter 1990): 106, 112 (italics in original), 116–17.

5. On the possible relationships between history and philosophy, see, for example, Paul Durbin, "History and Philosophy of Technology: Tensions and Complementarities," in *In Context: History and the History of Technology*, ed. Cutcliffe and Post. For examples of the conceptual borrowing among sociologists and historians of technology, see Wiebe E. Bijker, Thomas P. Hughes, and Trevor Pinch, eds., *The Social Construction of Technological Systems: New Directions in the Sociology and History of Technology* (Cambridge: MIT Press, 1987), passim, but especially 10–13, 51–52, 84, 112. Thomas P. Hughes, *Networks of Power: Electrification in Western Society, 1880–1930* (Baltimore: Johns Hopkins University Press, 1983). Andrew Pickering, in his book, *The Mangle of Practice: Time, Agency, and Science* (Chicago: University of Chicago Press, 1995), 217n.4, has argued that his earlier edited collection of essays, *Science as Practice and Culture* (Chicago: University of Chicago Press, 1992), which included contributions by philosophers, historians, sociologists, and anthropologists, "instantiates an antidisciplinary synthesis."

6. Weibe Bijker and John Law, eds. *Shaping Technology/Building Society: Studies in Sociotechnical Change* (Cambridge: MIT Press, 1992). John Staudenmaier's review, "Problematic Stimulation: Historians and Sociologists Constructing Technology Studies," is in *Research in Philosophy and Technology* 15 (1995): 93–102. My comments here are drawn from my own review of the latter volume in *Technology and Culture* 39 (January 1998): 122–24.

7. For examples of good undergraduate STS textbooks, see Rudi Volti, *Society and Technological Change,* 3d ed. (New York: St. Martin's Press, 1995) and Martin Bridgstock et al., *Science, Technology and Society: An Introduction* (Cambridge: Cambridge University Press, 1998).

8. Wil Lepowski, "Scholars Ponder the Goals of Science, Technology and Society Programs," *Chemical and Engineering News* 67 (1989): 14–15.

9. Susan E. Cozzens, "The Disappearing Disciplines of STS," *Bulletin of Science, Technology & Society* 10 (1990): 1–5. Cozzens continues to espouse this view of STS, and her essay has been reprinted in slightly updated form in Stephen H. Cutcliffe and Carl Mitcham, eds., *Visions of STS: Contextualizing Science, Technology, and Society Studies* (Albany: SUNY Press, 2000).

10. Paul T. Durbin, "Defining STS: Can We Reach Consensus?" *Bulletin of Science, Technology & Society* 11 (1991): 187–90. See also his more extended work, *Social Responsibility in Science, Technology, and Medicine* (Bethlehem, Pa.: Lehigh University Press, 1992), which I discuss in some depth in chapter 5.

11. Wiebe Bijker, "Do Not Despair: There Is Life after Constructivism," *Science, Technology, & Human Values* 18 (1993): 113–38.

12. Wiebe Bijker, "Understanding Technological Culture through a Constructivist View of Science, Technology, and Society," in *Visions of STS*, ed. Cutcliffe and Mitcham. See also Bijker's book-length study, *Of Bicycles, Bakelites and Bulbs: Toward a Theory of Sociotechnical Change, Inside Technology* (Cambridge: MIT Press, 1995), especially chapters 1 and 5, where he expands upon these ideas in greater detail. On the dynamic, interactive relationship between society and technoscience, see also two essays by Steven L. Goldman, "The Technē of Philosophy and the Philosophy of Technology," *Research in Philosophy and Technology* 7 (1984): 115–44 and "The Social Captivity of Engineering," in *Critical Perspectives on Engineering and Applied Science*, ed. Paul T. Durbin, Research in Technology Studies, vol. 4 (Bethlehem, Pa.: Lehigh University Press, 1991), 121–45.

13. I have drawn upon Yves Gingras, "Following Scientists through Society? Yes, but at Arm's Length!" in *Scientific Practice: Theories and Stories of Doing Physics*, ed. Jed Z. Buchwald (Chicago: University of Chicago Press, 1995) for the cake image, and upon my own "The Warp and Woof of Science and Technology Studies in the United States," *Education* 113 (Spring 1993), especially 381–82, for the weaving image.

14. Bijker, "Understanding Technological Culture"; Langdon Winner, "Technologies as Forms of Life," *The Whale and the Reactor* (Chicago: University of Chicago Press, 1986), chapter 1.

15. Carl Mitcham has suggested, for example, that what may be most important in any case are not methodologies per se, but rather the problems or realities engaged, and in this sense STS can be seen "as part of a broad cultural search for unities, relationship, meaning." Carl Mitcham, "Interdisciplinary Technological Research and STS Programs in the USA," unpublished conference paper from "Interdisziplinare Technikforschung und Ingenieurausbildung: Konzepte und Erfahrungen in verschiedenen Landern," June 18–20, 1991, Weingarten bei Karlsruhe. On the issue of interdisciplinarity, see also Stephen Jay Kine, *Conceptual Foundations for Multidisciplinary Thinking* (Stanford, Calif.: Stanford University Press, 1995).

16. David Hess, *Science Studies: An Advanced Introduction* (New York: New York University Press, 1997), 155.

17. On this shift, see, for example, Steve Woolgar, "The Turn to Technology in Social Studies of Science," *Science, Technology, & Human Values* 16 (Winter 1991): 20–50.

18. See Hess, *Science Studies*, especially chapter 5, upon which I have drawn heavily in this section by way of organizational framework and for examples of this shift. It should be pointed out, in light of the previous section, that, in the main, Hess views STS as an interdisciplinary endeavor.

19. Jacques Ellul argues a similar point in *The Technological Society* (New York: Alfred A. Knopf, 1964), 363–75, when he talks about the role of propaganda and the power of the state to induce its citizens into readily accepting the beneficence of technique without readily considering the more problematic consequences. See also his more lengthy assessment of this theme, *Propaganda: The Formation of Men's Attitudes* (New York: Alfred A. Knopf, 1965).

20. Hess, *Science Studies*, 114–16. See Franz Foltz, *The Increasing Participation in Science: The Rise and Fall of the U.S. Global Change Research Program* (Bethlehem, Pa.: Lehigh University Press, forthcoming); Andrew Webster, *Science, Technology, and Society: New Directions* (New Brunswick, N.J.: Rutgers University Press, 1991), 134.

21. Hess, *Science Studies*, 114–15; Georg Lukacs, "Reification and the Consciousness of the Proletariat," in *History and Class Consciousness* (Cambridge: MIT Press, 1968); Donna Haraway, *Primate Visions: Gender, Race, and Nature in the World of Modern Science* (London: Routledge, 1989).

22. Hess, *Science Studies*, 119–20; Haraway, *Primate Visions*, 290; Ruth Hubbard, *The Politics of Women's Biology* (New Brunswick, N.J.: Rutgers University Press, 1990), 119, 138; Marianne Van den Wijngaard, *Reinventing the Sexes: The Biomedical Construction of Femininity and Masculinity*, Race, Gender, and Science Series (Bloomington: Indiana University Press, 1997), quotations 1, 120; and Dianne Hales, *Just Like a Woman: How Gender Science Is Redefining What Makes Us Female* (New York: Bantam, 1999), xiv. See also Nancy Tuana, ed., *Feminism and Science* (Bloomington: Indiana University Press, 1989).

23. Susan Griffin in L. Caldecott and S. Leland, eds., *Reclaim the Earth* (London: Women's Press, 1983) and *Women and Nature* (New York: Harper and Row, 1978); Hillary Rose, "Hand, Brain, and Heart: A Feminist Epistemology for the Natural Sciences," *Signs* 9 (Autumn 1983): 73–90.

24. Marilyn Strathern, "No Nature, No Culture: The Hagen Case," in *Nature, Culture, and Gender*, ed. C. MacCormack and Marilyn Strathern (Cambridge: Cambridge University Press, 1980); Sandra Harding, *The Science Question in Feminism* (Ithaca, N.Y.: Cornell University Press, 1986), chapter 6, quotation 129. See also L. Segal, *Is the Future Female? Troubled Thoughts on Contemporary Feminism* (London: Virago, 1987); and H. Eisenstein, *Contemporary Feminist Thought* (London: Allen and Unwin, 1984).

25. For a good summary discussion of the issue of essentialism in both science and technology studies, see Judy Wajcman, *Feminism Confronts Technology* (University Park: Pennsylvania State University Press, 1991), especially chapter 1. The standard work on household technology is Ruth Schwartz Cowan, *More Work for Mother: The Ironies of Household Technology from the Open Hearth to the Microwave* (New York: Basic Books, 1983), but see also her various essays related to this theme, including "The Consumption Junction," in Bijker et al., eds., *The Social Construction of Technological Systems*, 261–80; also useful is Katherine Jellison, *Entitled to Power: Farm Women and Technology, 1913–1963* (Chapel Hill: University of North Carolina Press, 1993). On reproductive technologies, see Robbie Davis-Floyd, *Birth as an American Rite of Passage* (Berkeley: University of California Press, 1992); and Marilyn Strathern, *Reproducing the Future: Anthropology, Kinship, and the New Reproductive Technologies* (New York: Routledge, 1992).

26. Haraway, *Primate Visions*; Robert Bullard, *Dumping on Dixie: Race, Class, and Environmental Equity* (Boulder, Colo.: Westview Press, 1990).

27. David J. Hess, *Science and Technology in a Multicultural World: The Cultural Politics of Facts and Artifacts* (New York: Columbia University Press, 1995), ix, 7–9; Sandra Harding, *Is Science Multicultural? Postcolonialisms, Feminisms, and Epistemologies* (Bloomington: Indiana University Press, 1998), 2–3, 18–20,

191. Harding proposed the notion of "strong objectivity" in *Whose Science? Whose Knowledge? Thinking from Women's Lives* (Ithaca, N.Y.: Cornell University Press, 1991).

28. Hess, *Science Studies*, 134–36; Karen Knorr-Cetina, *The Manufacture of Knowledge* (New York: Pergamon, 1981); Bruno Latour and Steve Woolgar, *Laboratory Life: The Social Construction of Scientific Facts*, 2d ed. (Princeton: Princeton University Press, 1986).

29. Sharon Traweek, *Beamtimes and Lifetimes* (Cambridge: Harvard University Press, 1988); "Border Crossings: Narrative Strategies in Science Studies and among Physicists in Tskuba City, Japan," in *Science as Practice and Culture*, ed. Pickering, 429–65; *Big Science in Japan* (forthcoming).

30. On this issue, see Steven Turner and Karen Sullenger, "Kuhn in the Classroom, Lakatos in the Lab: Science Educators Confront the Nature-of-Science Debate," *Science, Technology, & Society* 24 (Winter 1999): 5–30, who note that most of the science education reform movements of the 1960s and 1970s focused on reinvigorating a rigorous curriculum based on notions of science as scientists viewed it. There was little regard for technology or societal implications, and the emphasis was on recruiting future scientists, not on educating responsible future citizens.

31. Brian Wynne, "Public Understanding of Science," in *Handbook of Science and Technology Studies*, ed. Jasanoff et al., 361–88, quotation 275.

32. Wynne, "Public Understanding," 275–82. On the effects of the Chernobyl radiation, see Brian Wynne, "Misunderstood Misunderstandings: Social Identities and Public Uptake of Science," in *Misunderstanding Science? The Public Reconstruction of Science and Technology*, ed. Alan Irwin and Brian Wynne (Cambridge: Cambridge University Press, 1996); "Unruly Technology: Practical Rules, Impractical Discourses and Public Understanding," *Social Studies of Science* 18 (1988): 147–67; and "Sheepfarming after Chernobyl: A Case Study in Communicating Scientific Information," *Environment* 31 (1989) 10–15, 33–39. For the AIDS case, see Steven Epstein, *Impure Science: AIDS, Activism, and the Politics of Knowledge* (Los Angeles: University of California Press, 1996); and "The Construction of Lay Expertise: Aids Activism and the Forging of Credibility in the Reform of Clinical Trials," *Science, Technology, & Human Values* 20 (1995): 408–37. Paula Treichler, "How to Have Theory in an Epidemic: The Evolution of AIDS Treatment Activism," in *Technoculture: Cultural Politics*, vol. 3, ed. Constance Penley and Andrew Ross (Minneapolis: University of Minnesota Press, 1991) also speaks to the issue. I have drawn my summary below from the descriptions of events presented by Harry M. Collins and Trevor J. Pinch in *The Golem at Large: What You Should Know about Technology* (Cambridge: Cambridge University Press, 1998), chapters 6 and 7.

33. Epstein, *Impure Science*, 228 and quoted in Collins and Pinch, *The Golem at Large*, 141.

34. Hess, *Science Studies*, 145.

35. Alan Sokal and Jean Bricmont in *Fashionable Nonsense: Postmodern Intellectuals' Abuse of Science* (New York: Picador, 1998), 184n.240, have attributed the apparent first usage of the term "science wars" to Andrew Ross in an essay "Science Backlash on Technoskeptics," *The Nation* 261 (October 1995), 346.

36. Paul R. Gross and Norman Levitt, *Higher Superstition: The Academic Left*

and Its Quarrels with Science (Baltimore: Johns Hopkins University Press, 1994), quotations below from 3, 9, 43, 45, 48–49, 234, and 239.

37. On this point see Brian Martin's review, "Social Construction of an 'Attack on Science,'" *Social Studies of Science* 26 (1996): 161–73. Also perceptive is Steven L. Goldman's review in *Science, Technology & Society Curriculum Newsletter* 107 (Spring 1996): 17–19.

38. Gross and Levitt, *Higher Superstition*, 5, 57, 107, 126–32. Latour's Third Rule of Method appears originally in *Science in Action: How to Follow Scientists and Engineers through Society* (Cambridge: Harvard University Press 1985), 99, 258. Latour has subsequently backed away from the extreme relativism implied in this earlier work as suggested by his book *We Have Never Been Modern*, trans. Catherine Porter (Cambridge: Harvard University Press, 1993), see especially 5–6. Here he talks of a "double construction—science with society and society with science," in which "objects are real" yet "look so much like social actors that they cannot be reduced to the reality 'out there' invented by the philosophers of science." Latour also addresses this issue in his more recent *Pandora's Hope: Essays on the Reality of Science Studies* (Cambridge: Harvard University Press, 1999), especially chapter 1, entitled "Do You Believe in Reality?" to which his short answer is "But of course," and in the conclusion. Also see Martin, "Social Construction of an 'Attack on Science,'" 171n.14 for several critiques of Latour's earlier work.

39. Paul R. Gross, Norman Levitt, and Martin W. Lewis, eds., *The Flight from Science and Reason*, Annals of the NYAC, vol. 775 (New York: New York Academy of Sciences, 1996), quotations, ix, 110–11.

40. Meera Nanda, "The Science Question in Postcolonial Feminism," in *The Flight from Science and Reason*, ed. Gross, Levitt, and Lewis, 429. For Donna Haraway's idea of "situated knowledge," see her "Situated Knowledges: The Science Question in Feminism and the Privilege of Partial Perspective," in *Simians, Cyborgs and Women: The Reinvention of Nature* (New York: Routledge, 1991).

41. Gross and Levitt, *Higher Superstition*, 100.

42. The special issue of *Social Text* 47/48 (1996) has been reprinted in slightly different form as Andrew Ross, ed., *Science Wars* (Durham, N.C.: Duke University Press, 1996) from which I have drawn my direct references. Quotations, 7–8.

43. Ross, "Introduction," in *Science Wars*, 11.

44. Langdon Winner, "The Gloves Come Off: Shattered Alliances in Science and Technology Studies," in *Science Wars*, ed. Ross, 102–5. An earlier version of this breakdown of STS can be found in Winner, "Conflicting Interests in Science and Technology Studies: Some Personal Reflections," *Technology in Society* 11 (1989): 433–38.

45. Sandra Harding, "Science Is 'Good to Think With,'" in *Science Wars*, ed. Ross, 17.

46. Alan D. Sokal, "Transgressing the Boundaries: Toward a Transformative Hermeneutics of Quantum Gravity," *Social Text* 46/47 (Spring/Summer 1996): 217–52; and "A Physicist Experiments with Cultural Studies," *Lingua Franca* 6 (May/June 1996): 62–64. It should be noted, however, that the essay utilizes numerous quotations taken directly from the works of selected cultural and science studies thinkers.

47. This characterization comes from Thomas Nagel's review of Sokal and

Jean Bricmont's subsequent volume, *Fashionable Nonsense: Postmodern Intellectuals' Abuse of Science* (New York: Picador, 1998), published in *The New Republic* (October 12, 1998): 32–38.

48. For a sampling of reactions to the hoax, see "Mystery Science Theater," a collection of responses including a somewhat self-justificatory essay by the editors of *Social Text*, a rejoinder by Sokal, and ten brief comments by scholars representing a wide range of disciplines, in *Lingua Franca* (July/August 1996): 54–64.

49. Alan D. Sokal and Jean Bricmont, *Impostures Intellectuelles* (Paris: Editions Odile Jacob, 1997). My quotations are from the English translation, *Fashionable Nonsense*, x–xi.

50. Sokal and Bricmont, *Fashionable Nonsense*, especially chapter 4, quotations 53, 55, 58, 67, 79–80, 98, 102. The Feyerabend quotations are taken from *Against Method* (London: New Left Books, 1975), 296–96.

51. Sokal, "A Physicist Experiments with Cultural Studies," 62.

52. See, for example, Hess's *Science Studies*, 35–36, 154 and Bruce V. Lewenstein's Op Ed response to the Sokal hoax in *The Chronicle of Higher Education* (June 21, 1996): B1–2, for their characterizations of where most STS scholars stand on the issues of scientific objectivity and relativism. For another view, see R. G. A. Dolby in *Uncertain Knowledge: An Image of Science for a Changing World* (Cambridge: Cambridge University Press, 1996), who advocates a "modest realism."

53. See, for example, Sokal and Bricmont, *Fashionable Nonsense*, 203 and Gross and Levitt, *Higher Superstition*, 2–4, 107–8, 157, 224.

54. For suggestions regarding this "normalization," see Thomas Hellstrom, "The Functional Integration of the 'Science Wars' in STS," *Technoscience* 12 (Winter 1999): 15; and Steve Fuller, "Whose Style? Whose Substance? Sokal vs. Latour at the LSE," *Technoscience* 11 (Fall 1998): 9.

55. Timothy Lenoir, ed., *Inscribing Science: Scientific Texts and the Materiality of Communication* (Stanford: Stanford University Press, 1998); Noretta Koertge, ed., *A House Built on Sand: Exposing Postmodernist Myths about Science* (New York: Oxford University Press, 1998). See also Ullica Segerstrale, ed., *Beyond the Science Wars: The Missing Discourse about Science and Society* (New Brunswick, N.J.: Rutgers University Press, 2000).

56. Stuart Bill Leslie, "Who's Talking? Who's Listening? Who Cares? Reestablishing a Conversation in STS," plenary address, National Association of Science, Technology and Society Annual Meeting, March 4, 1999, to be published in the *Bulletin of Science, Technology & Society*. Leslie also suggested that the exchanges between the Gross-Levitt-Sokal and the Andrew Ross-STS camps were but academic "skirmishes." In his mind the real "science wars" battles were played out in the more public arena of the debates over the plans for the original *Enola Gay* exhibit at the Smithsonian Institution's National Air and Space Museum, an exhibit that was subsequently scaled back due to outside pressures, and the negative reaction from the scientific community to that same institution's "Science in American Life" exhibit at the National Museum of American History. For a thoughtful response to the issues surrounding the *Enola Gay* incident, including reviews of four books published on the affair, see the special section "The Last Act: Reviewing Public History in Light of the *Enola Gay*," in *Technology and Culture* 39 (July 1998): 457–88, while comments on the "Science in American

Life" exhibit by two Smithsonian historians can be found in Robert C. Post and Arthur P. Molella, "The Call of Stories at the Smithsonian Institution: History of Technology and Science in Crisis," *ICON: Journal of the International Committee for the History of Technology* 3 (1997): 44–82.

57. Gross and Levitt, *Higher Superstition*, 239, 244, 246, 248, 253; Sokal and Bricmont, *Fashionable Nonsense*, 16, 185–86, 203. Here I have admittedly run together quotations from the two volumes for ease of reference, although I recognize there are some differences in viewpoint and tone between them. Nonetheless, together the authors do represent a fairly coherent overall position. In any case the specific references can easily be sorted out for anyone so wishing.

58. Sandra Harding, *The "Racial" Economy of Science: Toward a Democratic Future* (Bloomington: Indiana University Press, 1993), ix, 16–19.

59. Sandra Harding, *Is Science Multicultural? Postcolonialisms, Feminisms, and Epistemologies* (Bloomington: Indiana University Press, 1998), 19–20, 124–29, 190–91. Harding's *Whose Science? Whose Knowledge?* (Ithaca, N.Y.: Cornell University Press), especially chapter 6, has a full account of her idea of "strong objectivity." To be fair, it should be noted that some STS scholars such as David Hess, *Science Studies*, 46–47, have suggested that Harding's "strong objectivity" and standpoint epistemologies, which have only begun to coalesce, may be better suited to some fields like the social, environmental, or biomedical sciences, where social values may have had more epistemological impact, than in, say, theoretical physics. In any case, there is a need to extend such analyses into broader social scientific theory. Harding would presumably not disagree, and, in fact, in many ways that is what she is attempting in *Is Science Multicultural?*

60. Gross and Levitt themselves suggest that some of the more "grotesque" "critiques of science," if allowed their "current ascendancy among humanist radicals," will create "a rigid barrier" between Snow's "two cultures," one "maintained by mutual disesteem between scientists and their would-be critics." *Higher Superstition*, 244. Of course, Gross and Levitt are partly correct in their assessment here, admittedly because there are some extreme radical voices, although this does not necessarily mean they should not be at least be heard, but more importantly because they themselves, in the misdirected fury of their self-appointed protection of the notion of scientific "objectivity," have missed the point that STS may be the one best chance there is of constructive dialog on the societal issues surrounding science and technology. Thus, by deriding STS, they themselves are only building higher the barrier, rather than closing the gap of understanding. C. P. Snow himself hinted at this very point in the second edition of the published edition of his now-famous talk, when he identified "a body of intellectual opinion," largely in the social sciences, "concerned with how human beings are living or have lived," which he observed "is becoming something like a third culture." *The Two Cultures and a Second Look* (Cambridge: Cambridge University Press, 1964), 69–71. Although it was too early to identify this third culture as that of STS, and it would be presumptuous to assume that Snow would approve of how it has evolved or of its messages, any more than have the critics of STS, his observation was generally accurate and prescient of the phenomenon that would grow into STS studies beginning roughly at the end of the 1960s. On the need for continued "temperate" dialog, see also Lewenstein, "Science and Society: The Continuing Value of Reasoned Debate."

4

STS Programs, Institutions, and Journals

For most of us, the critical perspective offered by STS will be a permanent feature of all liberal education, as long as we remain alive.

— David Edge, *Reinventing the Wheel*

A field of academic study can be said to have arrived and, in many ways, can be measured by the establishment of degree-granting academic programs and the creation of professional societies and journals as outlets for the exchange of scholarly and educational pursuits. During the past three decades, a number of such programs, organizations, and publishing outlets have emerged, each of which contributes in important ways to promoting the central focus of STS, that is, the analysis and explication of science and technology as complex "social constructs" entailing cultural, political, economic, and general theoretical questions.

As the STS field has evolved, at least two, if not three, somewhat distinct, even if overlapping, interdisciplinary "subcultures" have emerged and are, in some sense, vying to carry the STS mantle. This tripartite division would seem to hold true both for the United States and, albeit with important distinctions, for much of Europe as well. This chapter will sketch out a rough map of this STS institutional terrain, focusing primarily on developments in the United States and Europe, but where possible expanding to cover activities in other parts of the world as well.[1]

A Conceptual Framework of Analysis

Although STS has always had multiple foci, the theme of STS "subcultures" was first systematically explored by Juan Ilerbaig in an essay published in the *Science, Technology & Society Curriculum Newsletter* in which he described a split between more disciplinary, theory-oriented scholars, often led by European sociologists of science, and more interdisciplinary, issue-centered educators, commonly led by philosophers of technology and engineering ethicists. He further characterizes the dichotomy by attributing to the former a strong science orientation with a more descriptive approach, while noting the latter's technology focus accompanied by normative or evaluative approaches. In a prompt rejoinder in a subsequent issue of the same newsletter, philosopher Steve Fuller characterized the split as a "High Church–Low Church" distinction, a catchy turn of phrase that quickly caught on with some STS scholars. In this view Fuller recognized what he saw as an unfortunate division between those programs, often at the graduate level, with "a discipline-centered, scholarly bent" and those with "a problem-centered, social activist bent." Far better in his mind would be an STS *movement* that would at once meld the activist strains of STS with the body of sustained "critical" knowledge regarding science (and technology) generated by sociological scholars.[2]

Other scholars continued the discussion, including Leonard Waks, who emphasized the distinction between what he sees as the knowledge- and empirically oriented "academics" and the more "meliorist," or "activist," social movement educators. Waks would apparently add the historians of science and technology to the lists of the former, but Luis Pablo Martinez took issue with this assignment of historians in a thoughtful paper in which he argues for an "activist" role for historians of technology because of their ability to "contextual[ize] accounts of technological developments in the past." Although speaking to a different audience in his presidential address to the Society for the History of Technology, Alex Roland argued much the same point in rationalizing the value of the history of technology. He views the field as a community of scholars that has amassed a knowledge base essential to understanding how technology contributes to societal and contextual change.[3]

Still other scholars have pushed the debate even further. Li Bocong, a philosopher in the Department of Science and Technology at the Chinese Academy of Sciences, has called our attention to the cultural split between already developed, even postindustrial, nations and those such as China still in the process of industrializing, and the implications this has for the STS field. Richard Gosden of the Department of Science and Technology Studies at the University of Wollongong in an essay in *Technoscience*, the newsletter of the Society for the Social Studies of Science, sees the High Church–Low Church distinction, which he characterizes as being "principally oriented in their research either to the problem of 'truth,' or alternatively, to the problem of 'justice,'" as being further fragmented into what he views as four "corner posts" for the field. He identifies these posts as:

1. the dominant form of "justice" within our society, that is capitalism or market justice (MJ);
2. its catchall alternative of victim justice (VJ);
3. the dominant epistemological authority within our society of scientific positivism (SP); and
4. its epistemological antagonist—science-as-social-construction, scientific relativism (SR).

Gosden accepts that this depiction of the STS field is overly neat and subject to further change as the boundaries continue to readjust themselves.[4]

Philosopher of technology Carl Mitcham has, in similar fashion to Gosden, depicted a matrix of four alternative approaches to STS in theory and practice. On one axis he breaks STS down into an academic field on one end and as a social movement on the other, while on the second axis the division is between those who are supportive of technoscience and those who are critical of its societal implications. Thus, the STS social action movement, on the one hand as a form of protest, "vocally questions whether the development of technoscience is always beneficial to society," while, as technological management on the other hand, it "aspires to infuse the management of science and technology with more consciously focused policy analysis and more thoroughgoing rational administration." Among academic programs there is a similar sort of split among those that tend to critique the technocratic society and those that "seek to instill the new technoscientific society with a deeper public understanding of the science and technology on which it relies," so that citizens can be "active, intelligent participants in social decisions that affect their lives."[5]

I have argued that, in addition to the High Church–Low Church distinction among STS programs, often characterized as Science and Technology Studies (S&TS) and Science, Technology, and Society (STS) respectively, there is a third approach often referred to as Science, Technology, and Public Policy (STPP) or sometimes Science, Engineering, and Public Policy (SEPP). The first two are oriented toward the theoretical/explanatory and the social/activist respectively. In contrast, STPP programs take a professional orientation with a focus on analyses of large-scale sociotechnical interactions and their management. They stress the need for, and training in, appropriate policy and management fields. Independent of whether one conceptualizes STS in terms of varied steeple heights (Fuller), as a three-legged tripod (Cutcliffe), or as a four-cornered field bounded by dueling "whipping posts" (Gosden), I believe it is fair to say that there are a variety of approaches to STS, many of which are admittedly overlapping and not necessarily mutually exclusive. In fact, it should be noted that, despite their differences, each of the authors noted in this discussion would like to see some sort of "rapprochement" (Waks's term) among the varied groups that would lead to a more truly interdisciplinary, if not transdisciplinary, understanding and treatment of the science-technology-society matrix. I believe that each approach has something valuable to contribute to the discussion. Recognizing then their nonexclusivity, and their value to understanding science and technology within their societal context, let me turn now to a discussion of what I view as the three main approaches to STS, drawing on a few selected illustrative examples.[6]

Science, Engineering, and Public Policy Programs

Professionally oriented STPP/SEPP programs stress the need for, and training in, appropriate policy and management fields. The most recent, third edition of the American Association for the Advancement of Science guide to such programs, edited by Albert Teich, identifies over forty such graduate-level programs both within the United States and around the world.[7] Typically these programs have a strong scientific and technical focus and are designed to train scientific and engineering managers in the broader sociopolitical context they are likely to encounter, or they have a more explicit administrative focus with the intent of training policy specialists. This area emerged in the late 1960s out of the concerns of engineers and technology managers and became institutionalized in the 1970s. Eventually these programs became closely allied with economists and managerial/policy professionals as well as engineers. STPP/SEPP programs are focused primarily on practical, career-oriented graduate training and are perhaps the largest and most well-developed group at that level. Three major journals representative of this approach to STS are *Technology in Society*, which is associated with the Society for Macroengineering; *Issues in Science and Technology*, a joint publication of the U.S. National Academy of Science, National Academy of Engineering, and the Institute of Medicine; and *Research Policy*, an international journal with a strong technology research, management, and policy orientation, with editorial offices at the Science Policy Research Unit at the University of Sussex, a program described briefly below.

Included among the roughly two dozen such programs currently active in the United States are the MIT Technology and Policy Program (1976), the Washington University of St. Louis Department of Engineering and Policy (1971), and the Carnegie Mellon Engineering and Public Policy Program (1970).[8] Each of these departments offers several degrees or tracks with varying emphases among engineering, management, or policy. With the exception of a masters track in Technology and Human Affairs at Washington University, each of the programs requires its entering students to have undergraduate or M.S. degrees in an engineering, science, or mathematics. The intent, as stated by Carnegie Mellon, is to "draw upon the skills and tools characteristic of various technical and social science disciplines" and to apply them "to technology and public policy problems, preparing engineers and scientists for work in both the public and private sectors." Graduates find employment in a wide range of corporate, governmental, and consulting opportunities, but generally with a technical orientation of some sort. It should be noted that there are other more social science-oriented SEPP programs, such as that offered by George Washington University (1970), in which the entrance requirements and program focus are less technically oriented, focusing instead on public policy perspectives. Another interesting program in this regard is that of the Center for Energy and Environment at the University of Delaware, which offers a Ph.D. degree and where many of its graduate students pursue research related to issues of sustainable development.[9] Additionally, graduate programs that more naturally fall into the S&TS category, such as Cornell University described below, may well also have a strong policy component. Thus, it should be clear that there can be no definitive description of what constitutes a STPP/SEPP program. Indeed, the same department or program often includes a range of program tracks, and much depends on the individual students' and faculty members' interests as to the emphasis taken within a given program.

In Europe there exists a parallel, but somewhat differently focused, policy and management approach to STS. There the emphasis would appear to be more heavily weighted to technical management, economics, and innovation studies than in the United States.[10] In France, the Conservatoire National des Arts et Métiers (CNAM) offers degrees in science policy and the economics of R&D and of technical development, while the Ecole Centrale Paris offers an M.S. in Technology Management for nonengineers, intended for students from business, economics, or political science backgrounds. The latter also offers a Ph.D. in Management of Technology and Innovation. It should be noted that CNAM is also the only institution in France that offers a Ph.D. in Science, Technology, and Society, suggesting that, as with many U.S. universities, a given institution may have several different program or degree tracks. In England the Science Policy Research Unit of the University of Sussex offers several different masters level degree tracks and one D.Phil. program, all of which have a heavy focus on "understanding complex and multifaceted 'real world' problems" with innovation management, development, and policy formation and implementation orientations. In Denmark, Roskilde University's Program in Technology Policy, Innovation, Socio-Economic Development (1988) offers a Ph.D. in Innovations Studies. Its express purpose is "to provide a theoretical as well as a more policy-oriented understanding of innovation in the context of industrial development and national welfare."[11] The University of Twente, a technical institution in the Netherlands, through its School of Philosophy and Social Sciences, began a combined engineering-philosophy masters degree program in Philosophy of Science, Technology, and Society (1983), "which trains students to become 'philosophical engineers.'" Here the objective is to integrate engineering study with philosophy, technology dynamics, and the history of science in a four-year program. Students typically go on to work in corporate product development, consultancy work, or the policy arena, all places that are at "the crossroads of technology and society."[12]

An interesting and revealing example of a non-Western nation that has committed extensive attention to STS is that of China. At numerous Chinese universities, as well as at the Graduate School of the Chinese Academy of Science and the Chinese Academy of Social Sciences, there are whole departments and programs devoted to STS research and education. Here one finds that the boundaries between academic research, teaching, and economic policy and planning are somewhat blurred, but it is a good example of mixing STS theory with an applied focus.[13]

One important center of Chinese STS studies is the Institute of Science, Technology, and Society at Tsinghua University. The institute was established in the mid-1980s in response to a recognition of the impact of new technologies on Chinese society. The institute has some dozen professors, several of whom have studied abroad in Europe and the United States, and it offers both undergraduate and graduate degree programs. Approximately 500 undergraduates and 900 graduate students enroll in their courses each year. Among their offerings are standard history, sociology, philosophy, and policy studies of science and technology. In addition, there is a strong emphasis on development strategies, systems engineering, and regional planning. The emphasis on these policy areas is also strongly reflected in their scholarly research, which is frequently focused both on theoretical strategies and models for science and technology policy and planning, and on practical case studies coordinated with city, regional, or provincial governmental commissions and agencies. The Tsinghua Institute of STS has played a constructive role in raising the level of scientific and technical decision making in local and regional governments, and it has received several awards and prizes for its efforts. Much of this experience finds its way back into classroom teaching, thereby insuring a problem-oriented and applied focus to even the more theoretical aspects of the curriculum.[14]

Collectively, these programs, especially in the United States and in Europe, have a graduate training focus in areas directly related to technology management (and to some degree assessment), development, and innovation studies or science and technology policy. U.S. STPP/SEPP programs were generally founded earlier than similar programs in Europe and England, and there would appear to be more programs devoted to science and technology policy, whereas in Europe the emphasis is more heavily on technology management and innovations studies. However, U.S. technology management programs would appear to emphasize the technical/engineering preparation of its students to a greater degree than their European counterparts, although this is not an absolute distinction, as witnessed by Twente University's focus on the "philosophical engineer." To generally characterize such programs, it would be appropriate to place them in Mitcham's STS matrix as that part of the social movement supportive of technoscientific literacy and a more rational management of science and technology and its policy. Collectively this is a well developed and seemingly growing area of STS studies as measured both by the number of programs and by student enrollments.

Science and Technology Studies Programs

In contrast to the more practical career-oriented focus of the STPP/SEPP approach, Science and Technology Studies (S&TS) involves more theoretical investigations into the social and cultural context of science and technology and their functioning as social processes. Here the interest is primarily explanatory and interpretative. It grew out of the 1960s debates among historians, sociologists, and philosophers regarding the inadequacies of internalist-oriented accounts of the nature, origins, development, and funding of science and technology. Whether the approach is that of the social constructivists, the relativistic or so-called Strong Program in the sociology of scientific knowledge, or the contextual history of technology, the inclination has been to view science and technology from a broader perspective than through a single disciplinary-bound window. Largely dominated by sociologists, and to a somewhat lesser degree by anthropologists and policy analysts, S&TS has had its greatest contribution in the area of conducting detailed empirical cases studies and subsequently theorizing about the "socially constructed" nature of scientific knowledge and technical development. Thus, for the most part, Science and Technology Studies programs are located at the graduate level and are research-oriented, although to be sure many of these programs also engage in undergraduate general education as well.

Among the leading S&TS scholarly journals are *Social Studies of Science*, edited by David Edge in association with the Science Studies Unit at the University of Edinburgh, and *Science, Technology, & Human Values* (*STHV*), now published by the Society for the Social Studies of Science (4S) and edited by Ellswurth Furman of the Department of Sociology in association with the Science and Technology Studies Program at Virginia Polytechnic Institute and State University. Together these two journals mirror the growth and increased professionalization that has come to characterize the STS field. To peruse back issues of *STHV* and *Social Studies of Science* is to chronicle much of the development of STS. 4S, a relative newcomer to the field, having been founded only in 1975, now encompasses not only sociologists of science and technology, but also anthropologists, policy analysts, and some historians and philosophers as well, testifying to the interdisciplinary nature of STS.[15]

Truly interdisciplinary S&TS programs, especially those offering a Ph.D., are relatively new and few in number, yet they are suggestive of the institutionalization of the field in the 1980s.[16] Among the better known and well established programs in the United States can be found Cornell University's Department of Science & Technology Studies (1969, 1991), Rensselaer Polytechnic Institute's Department of Science and Technology Studies (1982), and Virginia Polytechnic and State University's Graduate Program in Science and Technology Studies (1986). Although not huge by comparison with many disciplinary departments at American state-funded universities, each of these programs does have a fairly large number (sixteen–eighteen) of core faculty drawn primarily, but not exclusively, from the major disciplinary perspectives on science and technology—history, philosophy, sociology, and politics. In addition, faculty from other departments and disciplines often participate, giving these programs an even broader interdisciplinary flavor.[17]

Graduate students are generally required to take introductory seminar sequences and methodology courses that provide them with an interdisciplinary background mix of theory and methods before they advance to focus on specialized research topics of their own design and choosing, which often combine approaches from several different fields. Ph.D. students from these S&TS programs would generally anticipate finding employment within the academic community as faculty members. It is somewhat early to know with what success they will meet as the numbers of Ph.D. degrees awarded to date has been fairly small. Anecdotal evidence suggests that economically induced "downsizing" within American universities has made an already difficult job market for traditionally trained doctorates even more difficult for interdisciplinary STS Ph.D.'s. Students earning masters level degrees have been generally successful in finding employment in a wide variety of positions including policy work, environmental organizations, journalism, and government.[18]

As with the STPP/SEPP approach, strong European parallels with developments in the United States exist, albeit again with important differences. In fact, one of the major differences is the relative earlier establishment of several of the European S&TS programs, in particular the Science Studies Unit at the University of Edinburgh (1966). Drawing on a strong base in the sociology of science, the University of Edinburgh offers both an M.Sc. in Science and Technology Studies and a Doctoral Program of Social and Economic Research on Technology that draws on the expertise of faculty both from the Social Studies Unit but other social science faculty as well. The latter program is designed not only for STS students, but "technologists concerned to examine wider social dimensions of their work" as well. As noted above, CNAM's Centre Science,

Technologie et Société (1978) is the only French institution granting a Ph.D. in STS. Most if not all of its graduate students are already professionally employed, with a third each coming from academia, industry, and government. Among the research activities pursued by faculty, many of whom serve as government and public interest sector consultants, and by students, are risk assessment, technology transfer and development, and technology management. The Department of Science and Technology Dynamics (1982) of the University of Amsterdam offers an interdisciplinary M.Sc. degree in Science and Technology Studies and a Ph.D. in Science Dynamics in conjunction with other departments in Philosophy, Sociology, Chemistry, and Biology. As with many introductory U.S. S&TS programs, it requires introductory science studies literature and methodology courses designed to expose students to the interdisciplinary nature of the field before they begin pursuit of individualized research projects. In Germany, the Science and Technology Policy Unit of the Department of Sociology, cooperating with the interdisciplinary Institute for Science and Technology Studies of the University of Bielefeld, offer both an M.A. degree in STS and a doctoral program—Genesis, Structure and Impact of Science and Technology—that brings together sociologists, historians, and philosophers of science. In Austria, the Institute for Theory and Social Studies of Science at the University of Vienna (1987) consists of two main units—the first dealing primarily with the philosophy and history of science, and the second focusing more explicitly on science and technology studies. Several universities in Spain, including the Polytechnic University of Valencia and the University of Barcelona through its Departamento de Logica, Historia y Filosofia de la Ciencia, offer graduate training in S&TS. The Spanish graduate programs seem somewhat less structured and more individualized, being dependent upon particular faculty interests. For most of the European programs, as should be evident from these brief descriptions, the line between S&TS and STPP/SEPP programs is somewhat permeable.[19]

As a general matter, the European S&TS programs tend to have smaller faculties (five–ten) and smaller numbers of graduate students at any given point in time in contrast to their American counterparts. However, perhaps in response to these size limitations, but certainly as part of the European Union's cooperative orientation, there are far more national and international academic exchange programs than in the United States. For example, perhaps the most extensive and ambitious is the Society, Science, and Technology in Europe M.A. program involving eleven

EU universities,[20] in which students spend the first semester at their home institution focused on core STS concepts and methodologies. Then in their second semester, they can move to an alternate participating institution to take advantage of specialty research strengths offered there. In addition, the EU-funded ERASMUS program allows doctoral students from more than a dozen participating consortia members to collaborate and exchange students.[21]

On a national level, one finds cooperative S&TS networks and joint programs as well. For example, in Spain, INVESCIT (Instituto de Investigaciones Sobre Ciencia y Tecnología—1985), is a consortia grouping some half-dozen universities including the Polytechnic University of Valencia, the University of Barcelona, the University of Oviedo, and the University of the Basque Country. It is designed to pursue interdisciplinary STS research that pushes beyond the traditional positivist approaches to the study of science and technology typical of the Franco era and to view them instead as the product of social, economic, and political processes.[22] In Denmark, there is a Ph.D. network of cooperating departments and universities, including Roskilde University, Aalborg University, Denmark's Technical University, and the Copenhagen Business School, that jointly offer graduate level course.[23] In the Netherlands, there is also a Ph.D. training network, the "Netherlands Graduate School in Science and Technology Studies," that links together the Department of Science Dynamics at the University of Amsterdam, the Science Studies research group at the University of Limburg, which offers a degree course in "Culture and Science Studies," the Departments of Philosophy of Science and Technology and of General Philosophy at the University of Twente, and the Working Party for Science Studies at the University of Groningen. Since 1986 the network has offered summer school programs and several multiday workshops during the course of each year.[24] A fourth such example, although somewhat more disciplinary-oriented, can be seen in the London Centre for the History of Science, Medicine and Technology (1987). The Centre combines the strengths and resources of the Centre for the History of Science, Technology, and Medicine, Imperial College; the Department of History, Philosophy and Communication of Science, University College London; and the Wellcome Institute for the History of Medicine; with close ties to the London Science Museum, to offer an M.Sc. degree. Many students come from educational or museum backgrounds and then go on to earn Ph.D. degrees. The program includes several historical emphases but also includes some philosophy training.[25] Taken together, the European S&TS graduate level programs suggest at once strong parallels with equivalent American programs, while at the same time showing distinct traits in terms of generally smaller size complemented by a more open willingness to share and exchange resources, faculty, and students than would normally be the case in the United States.[26]

Science, Technology, and Society

The third STS approach, Science, Technology, and Society, per se, arose out of the late 1960s and early 1970s concerns regarding needed changes in undergraduate education. Such STS programs and courses emphasize general education for intelligent responsible citizenship in a highly scientific-technological society. As such they can stress scientific/technical literacy for practical citizenship and/or the contextual analysis of science and technology as an end in itself. One of the first such programs, interestingly enough, was the Program on Science, Technology and Society at Cornell University, which appeared in 1969 at least in part as a response to campus unrest and the need to develop "interdisciplinary courses at the undergraduate level on topics relevant to the world's problems." A similar impetus for STS programs on campuses with engineering programs such as Lehigh University (1972) and MIT (1977), was the perceived need to "create educational experiences which bring humanistic perspective to the application and evaluation of technology," and "to explore the influence of social, political and cultural forces on science and technology, and to examine the impact of technologies and scientific ideas on people's lives."[27]

Although all three of these programs have gone on to develop graduate level emphases of one variety or another, they continue to have a much broader "Low Church" general undergraduate education mission as well. Initially conceptualized as programs designed to polish the "coarse" surface of an engineering student's technical education by adding a cultural veneer, such programs quickly attracted the interest and attention of a much broader segment of the undergraduate student population. Such an appeal was felt at a large number of other colleges and universities as well. Thus, by way of example, at the predominantly undergraduate level, Vassar College has a Science, Technology, and Society Program offering an interdisciplinary B.A. in STS, and Carleton

College offers an Environmental and Technology Studies concentration for B.A. and B.S. students, while Stanford University's Program in Science, Technology, and Society offers both B.A. and B.S. majors in STS to its undergraduates, and Pennsylvania State University's STS Program (1969), which grew out of a "Two Cultures Dialog" with influence from Cornell University, offers an STS minor. Colby College offers an STS minor that focuses on the historical and social dimensions of science and technology, which is intended to serve as "an interdisciplinary bridge" between the academic cultures of the humanities and the sciences, both natural and social. Duke University's undergraduate curriculum requires all students to take two STS courses as part of their general education requirements. All these programs and their faculty, as well as the students, recognize the "problematic" nature of science and technology for contemporary society. Of interest were and are such issues as work and leisure in a mechanized age, loss of privacy, nuclear weapons and power, computerization, and a wide variety of environmental and energy related themes including the issue of sustainable development. Institutionalized primarily during the 1970s, this STS approach can now be found in some 100 formal programs and at the individual course level at perhaps 1,000 additional U.S. colleges and universities.[28]

Similar to American universities with engineering programs has been the experience emerging out of the "idealism of the sixties" at the technically oriented University of Twente. In addition to its masters degree program discussed above, the School of Philosophy, in an attempt to make science and technology "society-relevant," utilizes an STS approach to insure that its engineering students understand "the internal and external functioning of organization," gain "insight in[to] the complexity of social reality," and are stimulated by "critical reflection on problems concerning technical and societal developments."[29] Similarly, the engineering-focused CNAM Centre Science, Technologie et Société enrolls some 500 undergraduate students, including 250 majors. As at many U.S. programs with an S&TS focus, European S&TS-oriented programs also have strong undergraduate course offerings. Thus, the University of Amsterdam's Department of Science Dynamics annually enrolls more than 300 students in STS-related courses, while the University of East London's Department of Innovation Studies has more than 200, approximately half of whom are majors or minors.[30] One final example of STS-type offerings can be found at the University of Barcelona, where several hundred students a year take courses.

One major difference that appears to distinguish European STS from that found in the United States is that there are few four-year undergraduate-only colleges. It is likely that if a European university department or program offers STS courses at all, they will do so both on the undergraduate level and in some fashion at the graduate level as well. Thus, in Britain and Europe there appear to be few undergraduate-only STS programs and course offerings, in contrast to what is a fairly widespread phenomenon in the United States.

Typifying this curricular approach are two important publications, the *Science, Technology & Society Curriculum Newsletter*, published under the auspices of the Lehigh STS Program since 1977, and the *Bulletin of Science, Technology & Society*, established in 1981 and long associated with the Pennsylvania State University program, although it is now published by Sage. The *Bulletin* is also associated with the National Association of Science, Technology, and Society (NASTS, established in 1988), and beginning in 2000 is edited by Willem H. Vanderberg, Director of the Center for Technology and Social Development at the University of Toronto.

NASTS was established to create an interdisciplinary forum, in particular through an annual conference, its newsletter now entitled *STS Today* (formerly *NASTS News*), and the associated *Bulletin*, through which a wide range of individuals and groups expressing interest in, and concern for, the interrelationships among science, technology, and society could come together for the exchange of ideas and materials. NASTS consciously attempts to encompass a diverse community of individuals with broad ranging interests: kindergarten through twelfth-grade and postsecondary educators, policy makers, scientists and engineers, public interest advocates, science museum staff, religion professionals, print and broadcast media personnel, lay citizens, and representatives from the international community. Currently their interests are loosely grouped into four major areas of interest: education; ethical and sociocultural issues; science, technology, and public policy; and interdisciplinary scholarship. Of the four, education, and especially education at the kindergarten through twelfth-grade level, is clearly the strongest component of the membership, although all areas are certainly represented.

This broad based effort is the mark that distinguishes NASTS and thereby separates it from most other STS organizations; however, it also presents some problems for the organization. In particular, some members find NASTS's inclusiveness too broad, incorporating as it does individuals at one end of the spectrum who are highly critical of the science and technology "as usual" approach, and those at the other end of the spectrum who are willing to advance little criticism and wish primarily to improve and extend public understanding and acceptance, albeit often through enhanced and innovative pedagogies of science and technology. This spectrum of views was, for example, vividly portrayed through a series of exchanges occasioned by Chellis Glendenning's essay, "Notes Toward a Neo-Luddite Manifesto," and several equally pointed member responses published in *NASTS News*.[31]

As a broad based membership organization, NASTS has sought to provide appropriate forums through its annual meeting and publications wherein one can explicate the complexities of science and technology by noting their positive and negative impacts (both expected and unexpected), by analyzing how scientists and engineers go about their work, by scrutinizing the ways in which societal institutions contribute to the shaping and development of science and technology, and by suggesting better mechanisms for better control over the scientific and technological processes. It seeks to do so in a "constructively critical" way by providing a venue in which its members can meet, share concerns, and debate differing ideas on the nature of, and how best to handle, science and technology in today's society. To date it has been most successful in doing so in the educational arena.[32]

General Education and the Primary-Secondary Level

During the past two decades, STS in the context of general education, especially as it applies to primary and secondary, or K–12, education has become increasingly important. While most of the discussion about STS in a theoretical framework applies to general education at the K–12 level, there are some unique aspects, which create yet another subculture within STS. Some STS educators, perhaps most notably Rustum Roy, the primary founder of NASTS, argue strongly for the recognition of STS as lying at the core of general education. Roy, who has variously referred to STS as a "megatrend" and the "unsung revolution in education," also views STS as providing an alternative entrance point to, and perhaps the most effective form of, science education for the median student. In essence, he and others argue that the current science paradigm—Physics, Chemistry, Biology, which Roy jokingly refers to as PCBs—is good for only about 5 to 10 percent of American students. Furthermore, he argues, the current paradigm generally fails to talk about technology very much at all. In contrast, STS tries to make an interdisciplinary, real-world bridge among the disciplines as an interactive component at the very center of general education. This is important for all students, but especially the 90 to 95 percent who will likely not go on to careers in science and engineering.[33]

During the course of the 1980s, a series of educational studies and reports reflected on the status of American education, especially with regard to the roles played by science and technology. In general, they lamented school students' understanding of scientific principles, the relationships of those principles to technological application, and the societal context in which science and technology occur. In reaction to the generally agreed-upon failings of U.S. science education, a number of responses emerged, and among them the STS approach has increasingly come to be seen, at least in some quarters, as particularly viable.

Historian of science and technology Steven Turner and science educator Karen Sullenger in a recent article have described how science educators have drawn selectively on the nature-of-science literature as a resource for various reform efforts. They examine four curricular reform guides including the National Academy of Science's *National Science Education Standards* (1996), the Association for the Advancement of Science's Project 2061, the National Science Teachers Association's (NSTA) *Scope, Sequence and Coordination of Secondary School Science* (1992), and the Canadian Council of Ministers of Education's *Common Framework of Science Learning Outcomes* (1997). The NSTA's document is the most philosophically conservative with regard to science content and adopting a realist stance, but all four seek to promote scientific literacy, argue for some understanding of technology, especially in terms of preparing future responsible citizens, and emphasize the training of a competent workforce. All draw on the academic science studies literature to some extent.[34]

Turner and Sullenger note that science educators within these reform efforts tend to draw on science studies in three ways: the history and philosophy of science (HPS), STS, and constructivist pedagogy. The HPS pedagogical movement has tended to distance itself from cultural and contextual interpretations of science, leaning instead toward the more conservative "science-as-rational" view, even while they use concrete case studies to humanize science and demonstrate the nature of scientific reasoning and theory development. Turner and Sullenger argue that the STS-based school movement has had somewhat less direct interaction with science studies academics than has HPS; nonetheless, this group is particularly concerned about science and technology's impacts on society. They also note that there are tensions within the movement regarding appropriate goals, that is, whether the focus should be on sociocultural understandings of science and technology, on contemporary problems or policy issues, or on the training of future, societally responsible citizens. One major fear is that science content is frequently sacrificed in favor of social issues, while another is a concern that too close an association with the more relativistic and critical STS movement within the universities will taint the science literacy movement with neo-luddism.[35] Turner and Sullenger suggest that this may be the reason why some of the curricular reform projects have been somewhat cool towards STS. I will return to this issue below, but first I need to mention briefly the third way in which Turner and Sullenger see science educators drawing on the nature-of-science literature, that of "constructivist" learning theory.[36]

In contrast to the social constructivist approach of the Sociology of Scientific Knowledge (SSK) scholars, educators use the term to describe a process of learning in which students are encouraged to set out their preconceived ideas, then to clarify those conceptions in group discussion with other students or in conflict situations that reveal their adequacies/inadequacies, and finally to "construct" new ideas. This multifaceted and sometimes self-contradictory approach frequently draws on science studies models of research science and the scientific community in terms of "sociocultural practice" for which HPS supporters have criticized them as being "antirealist." Turner and Sullenger themselves warn that such constructivist approaches to learning may not fully reflect issues of power, status, and institutional constraints, among other things. Given these differing approaches to science education reform, Turner and Sullenger also conclude that, "Educators must not shrink from the hard negotiation over what will count as an understanding of science, or from their obligation to devise humane and socially responsible methods for promoting that understanding."[37]

Turner and Sullenger may have downplayed the NSTA's enthusiasm for STS, for as early as 1982 the largest organization for U.S. school science teachers adopted a position paper advocating an STS approach to science education to "develop scientifically literate individuals who understand how science, technology, and society influence one another and who are able to use this knowledge in their everyday decision making."

This approach was reaffirmed again in 1990 with an updated version of this position paper on STS.[38] Robert Yager, chair of the NSTA Task Force on STS Initiatives, notes that the organization has identified ten features characterizing the STS approach. They include:

- using student-identified problems with local interest and scientific and technical components as organizers for the course;
- using local resources (human and material) as original sources of scientific or technical information that can be used in problem resolution;
- involving students in seeking scientific or technical information that can be applied in solving real-life problems;
- extending science learning beyond the class period, the classroom, and the school;
- focusing on the impact of science and technology on each individual student;
- viewing science content as more than something that exists for students to master for tests;
- deemphasizing the mastery of science process skills that merely mimic skills used by practicing scientists;
- emphasizing career awareness—especially careers related to science and technology;
- providing opportunities for students to perform in citizenship roles as they attempt to answer questions about the natural world and to deal with problems they have identified; and
- demonstrating that science and technology are major factors that will impact the future.[39]

Proponents thus see STS as an approach to science teaching that responds to many of the fundamental problems of science education in the United States, including comprehending basic science concepts and processes, declining interest in science careers, stifling natural curiosity, and the relationship of classroom science to life and work outside the school. Dennis Cheek of the Rhode Island Department of Education argues that "the crisis in science education in America is not one of failure to know scientific concepts and principles. It is the failure to actively engage students in learning about the interactions of science, technology, and society as it [sic] affects their lives now and in the future."[41]

Despite NSTA's enthusiastic support for an STS approach to science education, it is important to note, as Turner and Sullenger have suggested, that not all science educators are in complete agreement. Many, such as Bill Aldridge, the former executive director of NSTA, believed instructing all students in "basic science," albeit "with hands-on experience with phenomena and practical applications is our best hope for better science education." Aldridge, who headed NSTA's Scope, Sequence, and Coordination of Secondary Science reform project, is wary of STS's "constructivist" learning approach. He and others fear STS students will fail to understand scientific models and theories if only exposed to the social, political, economic, and ethical aspects of technological applications. He argued STS needs clearer definition with a framework or core of basic science as a guide.[41]

Reader response to the "debate" between the two perspectives on STS reflected in the stances of Yager and Aldridge, which was presented in essay form to NSTA members, was somewhat split with twenty-eight members voting for the STS approach, nineteen opting for basic science, and fourteen falling into an "other" category. Although STS was the most strongly supported, many respondents noted the importance of not viewing this "debate" as an either/or argument.[42]

Another important development in secondary school science education with important ramifications for STS has been the American Association for the Advancement of Science's (AAAS) Project 2061. Taking its name from the date of the next return of Comet Halley, an event that most of the current generation of grade school children will live to witness, the AAAS science education reform effort includes a strong societal context component. For example, the history and nature of technology receive treatment in *Science for All Americans,* the project's initial summary of what students should know to be considered "scientifically literate," which was subsequently reaffirmed in its later *Benchmarks for Scientific Literacy*. Project 2061 is based on the assumption that the quality of life in that year will depend on the education and understanding of science and technology that this generation and the next will receive. Currently the project is working to help develop relevant curriculum materials.[43]

STS at the secondary level is probably most widely spread in terms of science education, a fact reflected in NASTS membership, of which the single largest group are school science educators. However, two other important groups, those of the social studies and technology educators, are also important components of the STS movement at this level.

Indicative of increased interest in STS issues on the part of social studies educators, has been the work of the National Council for Social Studies (NCSS) during the past decade or more.[44] Focused initially on energy- and science-related social issues, in 1990 the NCSS formally adopted a set of revised STS guidelines: "Teaching About Science, Technology and Society in Social Studies: Education for Citizenship in the 21st Century." The thrust of social studies STS education generally, and of the guidelines more specifically, has been on the education and development of responsible students with "an understanding of how science and technology shape and are shaped by society, the problems and opportunities they create, and how citizens can relate most effectively to them."[45]

The third major STS subgroup at the secondary school level is what has traditionally been referred to as the industrial arts, or more pejoratively, the vocational "shop" programs. Recognizing that traditional vocational education was inadequate to student needs in a rapidly changing workplace environment, "technology education" has recently developed in American secondary education. This change is reflected, for example, in the 1985 name change of the American Industrial Arts Association, founded in 1939, to the International Technology Education Association (ITEA), and their founding in 1989 of a new *Journal of Technology Education*. The intention of ITEA and other technology education groups is to broaden traditional vocational and technical *training* to include a broader *education* that incorporates the social impacts and context of technology generally, and more specifically of communications, transportation, and energy systems among others. One example of ITEA's efforts in this regard has been their effort to develop a set of curriculum standards for technology, parallel to those for science embodied in Project 2061. The project, which has received funding from the National Science Foundation and NASA, is entitled Technology for All Americans. The project's *Standards for Technology Education* has recently undergone a period of public review and discussion and is currently available for use. Among other foci, it stresses the need to recognize that "technology exists in the context of particular activities. . . . "[46] Technology educator Thomas Liao notes that New York State's recently published *Learning Standards for Mathematics, Science, and Technology* (MST) also calls for an integrated understanding in which students will seek to apply MST knowledge and skills in technology and society problem-solving activities.[47]

At the secondary school level then, there would appear to be three major subgroups contributing to STS education—science, technology, and social studies. The three strands, while clearly overlapping, are not completely interwoven; nonetheless, the thrust in all three areas is toward general education—providing students with an ability to see how society shapes science, and in turn how science and technology affect society and our individual values. Equally important to this intellectual understanding is the emphasis on how the student/citizen can effectively relate to science and technology and play a role in shaping and controlling them. This general education role of STS clearly places it within the third major interdisciplinary category or "subculture" noted above. Interestingly, however, where STS at the collegiate general education level has increasingly become institutionalized with its own set of departments, programs, and courses, this is much less so the case in the secondary schools. This is perhaps due to the more structured and traditional disciplinary divisions at the latter level. Whatever the reasons though, it has the advantage of spreading STS concerns and education across the major elements of the system, thus avoiding "ghettoization" or marginalization of STS. This is a not unimportant concern of any new, and especially any new *interdisciplinary*, field of study struggling for acceptance within a traditional educational framework.[48]

Conclusion

At the beginning of this chapter I suggested several different frameworks for thinking about the ways in which STS as a field of study is organized. They ranged from Steve Fuller's "High Church–Low Church" division to Carl Mitcham's four-part matrix in which he depicts STS as having a social movement side and an academic program component, each of which can be either promotional or protestingly critical of technoscience. My own view of the field is somewhat more akin to a tripartite division, broken down along the lines of "subcultures," each of which examines the relationship between ideas, machines, and values from a different perspective and experience. In some ways it is like the Japanese film story of *Rashomon*, in which the truth regarding the death of an individual is related entirely differently depending on the interests and perspective of the witness being interviewed. In my own view, a conjoining of all the perspectives is ultimately necessary.[49]

What holds all these STS subcultures together, despite their differences in approach and concerns, is a common appreciation for the complexities and contextual nature of science and technology in contemporary (and historical) society. In roughly three decades of STS research and curricular development, STS scholars and teachers have, for the most part, moved far away from internalist descriptive accounts of science and technology and from simplistic black and white, pro-con interpretations of their societal influences and interactions. Certainly it is true that "High Church" relativist scholars have frequently avoided passing prescriptive judgment regarding their case studies, while at times the more activist "Low Church" STSers have sounded particularly strident in expressing their views regarding the strengths or weaknesses of contemporary technoscience. It is also true that, as policy analysts or as STS educators, we have not done as much as we might have to convey to the general public at large and to our students enough about the societal complexities of contemporary science and technology. Resolving these individual failures, however, would not mean that all the subcultures would still not be necessary for the success of STS as an interdisciplinary field, for no one perspective, or subculture, by itself is sufficiently strong to bear the weight of leaning upon. Taken together, the three legs of the STS tripod are, to borrow a phrase from David Edge, in "creative tension," a tension that for the tripod to bear any weight must be seen and pursued as a source of vigor, not of division.[50] For only when we both understand science and technology in their broader societally-constructed context, and attempt to effectively shape them in response to societal goals, through the "critical perspective" noted by Edge at the head of this chapter, can STS scholars and educators claim any sense of closure. This is at once the hope for STS and its greatest opportunity, the topic to which I will turn in the following chapter.

Notes

1. One recent example of note has been the work of the Organizacion de Estados Iberoamericanos (OEI or Organization of Spanish-American States), which has been very active in the promotion of STS activities and education throughout Latin America. It has conducted a series of conferences and workshops and is helping to create specialized courses dealing with science and technology policy, science communication, and science education. While some of these courses are aimed at public administration officials in Latin American countries, efforts are also underway to develop and publish STS educational materials, to introduce STS into the core curricula at the secondary and tertiary levels, to establish an STS documentation center in Bogotá, and to create an STS-network. Further

information on this effort may be obtained electronically at: http://www.oei.es/cts.htm.

2. Juan Ilerbaig, "The Two STS Subcultures and the Sociological Revolution," *Science, Technology & Society Curriculum Newsletter* 90 (June 1992): 1–6; Steve Fuller, "STS as Social Movement: On the Purpose of Graduate Programs," *Science, Technology & Society Curriculum Newsletter* 91 (September 1992): 1–5. See also the further exchange between Ilerbaig, "On the Sociological Revolution in STS: A Fuller Account," and Fuller, "Give STS a Place on Which to Stand, and It Will Move the University—and Society," in *Science, Technology & Society Curriculum Newsletter* 92/93 (November/December 1992): 1–6; and Fuller's more extended treatment of these same issues in his book, *Philosophy, Rhetoric, and the End of Knowledge: The Coming of Science and Technology Studies* (Madison: University of Wisconsin Press, 1993). In his excellent recent introduction to the field, anthropologist David Hess also notes this dichotomous split between the early "science, technology, and society" activists and the more recent academic "professionalization" of "science and technology studies" that signaled what the former viewed as lessened concern for social justice. David J. Hess, *Science Studies: An Advanced Introduction* (New York: New York University Press, 1997), 2–3.

3. Leonard Waks, "STS as an Academic Field and Social Movement," *Technology in Society* 15 (1993): 399–408; Luis Pablo Martinez, "History of Technology and STS Studies: A Critical Approach," *Science, Technology & Society Curriculum Newsletter* 105 (Fall 1995): 1–7; Alex Roland, "What Hath Kranzberg Wrought? Or, Does the History of Technology Matter?" *Technology and Culture* 38 (July 1997): 697–713.

4. Li Bocong, "STS in China," *Science, Technology & Society Curriculum Newsletter* 104 (Summer 1995): 1–5; and Richard Gosden, "STS Whipping Posts Enclose the Discipline," *Technoscience* 8 (Fall 1995): 14–16.

5. Carl Mitcham, "Science-Technology-Society in Theory and Practice: A Conceptual Introduction," in *Para Comprender Ciencia, Tecnología y Sociedad (Spanish Language Handbook of Science, Technology and Society)*, ed. Andoni Alonzo et al. (Estella: Editorial Vervo Divino, 1996), 9–12, quotations from unpublished English-language version, 6–7.

6. Cutcliffe, "The Warp and Woof of Science and Technology Studies in the United States," *Education* 113 (Spring 1993): 381–91, 352. The preceding references (notes 2–5) only partially reflect the extent of the discussion, for example, Langdon Winner's depiction of four major strands within STS noted in the final section of chapter 3 above, but clearly the use of "subcultures" in this debate draws on and reflects distinctions suggested by C. P. Snow some thirty-five years ago. See Snow, *The Two Cultures: And a Second Look* (Cambridge: Cambridge University Press, 1964). See also Gary Bowden, "Coming of Age in STS: Some Methodological Musings," in *Handbook of Science and Technology Studies*, ed. Sheila Jasanoff et al. (Thousand Oaks, Calif.: Sage, 1995), 64–79.

7. Albert H. Teich, ed. *Guide to Graduate Education in Science, Engineering, and Public Policy*, 3d ed. (Washington, D.C.: Directorate for Science and Policy Programs, American Association for the Advancement of Science, 1995); see also Paul T. Durbin, "Technology Studies against the Background of Professionalization in American Higher Education," *Technology in Society* 11 (1989): 439–54.

8. Program information is derived from Teich, *Guide to Graduate Education*

in SEPP; Stephen H. Cutcliffe and Carl Mitcham, "Una descripción de los programas y la educación CTS universitaria en Los Estados Unidos" (A Profile of Collegiate-level STS Education and Programs in the United States), in *Superando fronteras: Estudios europeos de ciencia-tecnología-sociedad y evaluación de tecnologías*, ed. José Sanmartín and Imre Hronsky. Nueva Ciencia, 11 (Barcelona: Anthropos, 1994), 189–218; Mitcham and Cutcliffe, *STS Directory*, 2d ed. (University Park, Pa.: National Association of Science, Technology, and Society, 1996); and various program brochures. The year in parenthesis indicates the year of founding. Carnegie Mellon's Department of Engineering and Public Policy began as an undergraduate program and did not begin granting graduate degrees until 1977.

9. Program brochure, Center for Energy and Environmental Policy, University of Delaware, 1995.

10. This section is admittedly based on the availability of a somewhat scantier body of information, but the recent survey by Paul Wouters, Jan Annerstedt, and Loet Leydesdorff, *The European Guide to Science, Technology, and Innovation Studies* (Brussels: European Commission, 1999), which was unavailable at the time this chapter was initially drafted, generally confirms these initial impressions. This guide is available on line at <www.chem.uva.nl/stsguide>. The current third edition of Teich, *Guide to Graduate Education in SEPP* contains for the first time information on non-U.S. SEPP programs. Mitcham and Cutcliffe, *STS Directory* also contains some information on international STS programs. Valuable for Scandinavian developments is Lars Fuglsang, *Technology and New Institutions: A Comparison of Strategic Choices and Technology Studies in the United States, Denmark and Sweden* (Copenhagen: Academic Press, 1993). See also Arie Rip, "Science Studies in the Netherlands," Dominique Pestre, "Recent Development and Institutionalization of Social Studies of Science in France," Chris Caswell, "Illumination and Exploration: Social Studies of Science and Its Users in the 1990s," and Ulrike Felt, "Institutionalization and Options for Future Development of Science Studies in Austria," in *Social Studies of Science in an International Perspective*, ed. Ulrike Felt and Helga Nowotny, Proceedings of a Workshop, University of Vienna, January 13–14, 1995 (Vienna: Institute for Theory and Social Studies of Science, University of Vienna, 1995), 15–50.

11. Felt, "Institutionalization and Options," 128–30; Roskilde University Graduate Program in Innovation Studies Booklet.

12. Jaap Jelsma, "Integrated Training of Engineers for a Changing Society, or How to Breed the Philosophical Engineer?" Paper presented at the Congreso internacional: Tecnología, desarrollo sostenible y desequilibrios, Universidad Politecnica de Catalonia, Terrassa, Barcelona, December 14–16, 1995.

13. My general observations on the Chinese STS scene are based on two trips made in 1992 and 1998 at which times I visited some half-dozen STS programs and met with representatives from other institutions at several meetings and conferences. For a more extended discussion, see my essay, "Some Impressions of Science, Technology and Society Studies in China," *Technology in Society* 15 (1993): 243–51. Also useful is Yin Deng-xiang, "STS Related Education in China," *Science, Technology & Society Curriculum Newsletter* 85 (September 1991): 11.

14. For a full report on this program, see Wei Hongsen, "The Development of

Science, Technology and Society at Tsinghua University, Beijing, China," *Science, Technology & Society Curriculum Newsletter* 91 (September 1992): 6–8.

15. To be sure there are many other English-language STS-oriented scholarly journals, to say nothing of the numerous international publications. Several of the more important would include: *Configurations; Perspectives on Science; Philosophy and Technology*; *Public Understanding of Science; Research in Philosophy and Technology; Science and Engineering Ethics; Technology and Culture;* and *Technology in Society.* Even the education-oriented *Bulletin of Science, Technology & Society* contains scholarly research-based essays, as well as pedagogical pieces.

16. I am consciously excluding from this discussion the many excellent, but more disciplinary-oriented graduate programs, such as the History and Sociology of Science Program at the University of Pennsylvania. Much good STS research is conducted at such programs and numerous graduate students are well trained, but the educational/research thrust is generally much less explicitly interdisciplinary in focus.

17. See relevant program descriptions in Teich, *Guide to Graduate Education in SEPP*; and Cutcliffe and Mitcham, *STS Directory*.

18. Periodic e-mail exchanges within the Science and Technical Studies Discussion Group on the Internet reveal both frustrations at the lack of available academic opportunities for interdisciplinarily trained students and reports that at least some students are finding employment outside the traditional academic career that is rewarding. The question of balancing the number of STS programs training appropriate numbers of graduate students with available employment opportunities, both within and outside the academic community, is as of yet an unresolved issue, but one that will bear close watching. It will be important both for the success and viability of those programs, but also will have a bearing on the next generation of college and university professors who will, or will not as the case may be, teach future undergraduate students about the interplay between science, technology, and society. The spring 1999 newsletter of Cornell's Department of Science & Technology Studies suggests that its Ph.D. students have been reasonably successful in finding appropriate employment ranging from postdoctoral research fellowships and independent consultancies to traditional faculty appointments at a wide range of colleges and universities. Information from RPI's Department of Science and Technology Studies suggests similar success in student placement.

19. Teich, *Guide to Graduate Education in SEPP*; Mitcham and Cutcliffe, *STS Directory*; Felt, "Institutionalization and Options"; and STS program brochures from University of Edinburgh and University of Amsterdam.

20. This M.A. program is sponsored by ESST (The European Inter-university Association on Society, Science and Technology). Participating universities include: University of East London, University of Namur–Belgium, University of Strasbourg, Roskilde University, University of the Basque Country, Ecole Polytechnique Fédérale de Lausanne, Louvain-la-Neuve, University of Limburg, University of Madrid, University of Oslo, University of Bari.

21. Participating members include: Department of Science Dynamics, University of Amsterdam; University of Limburg and Maastricht Economic Research Institute on Innovation and Technology, Holland; Department of Science and

Technology Policy, University of Manchester; Centre de Sociologie de l'Invention, Ecole Nationale Supérieur des Mines de Paris; Centre Science, Technologie et Société, Conservatoire National des Arts et Métiers (CNAM), Paris; Université de Paris; Program of Science and Technology Studies, Universität Bielefeld, Germany; Social Studies of Science and Technology Unit, Universität Wien, Austria; Centre for Technology and Society, University of Trondheim, Norway; Centre for Science Studies, University of Göteborg, Sweden; Science Studies Unit, Edinburgh University; Faculty of Sociology, Universita La Sapienza, Rome; Program in Innovations Studies, Roskilde University, Denmark.

22. Manuel Medina and José Sanmartín, "A New Role for Philosophy and Technology Studies in Spain," *Technology in Society* 11 (1989): 447–55; José Sanmartín and José A. Lopez Cerezo, "CTS en Espana: Instituto de Investigaciones sobre Ciencia y Tecnología," in *Superando Fronteras: estudios europeos de Ciencia-Tecnología-Sociedad y evaluacion de tecnologías*, ed. Sanmartín and Imre Hronszky (Barcelona: Anthropos, 1994).

23. Innovation Studies Program brochure, Roskilde University.

24. Rip, "Science Studies in the Netherlands," 16–18, 24.

25. London Centre for the History of Science, Medicine, and Technology brochure.

26. To be fair, it should be pointed out that some American universities do have consortia arrangements, especially at the undergraduate level, either as part of state systems of higher education, where two-year branch campuses feed students to a centralized main campus for the final two years, or for purposes of cross-registration between institutions, but usually only on a course-by-course basis. Individual faculty (and programs) at times cooperate on specific research projects as well. For example, Terry Reynolds (Michigan Technological University) and I coproduced, under the auspices of the Society for the History of Technology, the second edition of *The Machine in the University* (Bethlehem, Pa.: STS Programs, Lehigh University and Michigan Technological University, 1987), a sample collection of STS course syllabi, while Carl Mitcham (Pennsylvania State University) and I coedited the *STS Directory* under the auspices of the National Association for Science, Technology, and Society (University Park, Pa.: STS Program, Pennsylvania State University, 1996), a guide to some fifty STS programs around the world. In both these cases the cooperation went beyond mere coauthorship in that the institutions involved shared in production- and distribution-related expenses. However, the formality and extent of some of the European national and international exchange programs at the graduate level described here is extremely rare in the United States.

27. Franklin A. Long, *First General Report, Cornell University Program on Science, Technology, and Society* (Ithaca, N.Y.: Cornell University, 1971), 22; Edward J. Gallagher, *Humanities Perspectives on Technology, Annual Report Year Five, 1976–1977* (Bethlehem, Pa.: Lehigh University, 1977), iii; *Program in Science, Technology and Society* (Cambridge, Mass.: MIT, 1980), 3.

28. See, for example, the admittedly incomplete, Mitcham and Cutcliffe, *STS Directory*; and Cutcliffe and Mitcham, "Una descripción de los programas y la educación CTS." A description of the Colby program can be found in "The Minor for All Majors: STS and the Liberal Arts at Colby College," *Bulletin of Science, Technology & Society* 18 (December 1998): 458–59. For Duke University's new

requirements, see Alison Schneider, "When Revising a Curriculum, Strategy May Trump Pedagogy: How Duke Pulled off an Overhaul while Rice Saw Its Plans Collapse," *The Chronicle of Higher Education* 45 (February 19, 1999): A14–16.

29. Jelsma, "Integrated Training of Engineers for a Changing Society," quotations, 4–5.

30. Mitcham and Cutcliffe, *STS Directory*.

31. Chellis Glendenning, "Notes Toward a Neo-Luddite Manifesto," *NASTS News* 3, no. 3 (1990); Irma Jarcho et al., "Letter to the Editor," *NASTS News* 3, no. 4 (1990): 6; Morris Shamos, "An Open Letter to Irma Jarcho, John Roeder, and Nancy Van Vraken," *NASTS News* 3, no. 6 (1990): 3–4; William F. Williams, "To Be Critical or Not to Be—That Is the Question," *NASTS News* 3, no. 6 (1990): 3–4; and E. H. Henninger, "Letter to the Editor," *NASTS News* 4, no. 1 (1991): 3, 10.

32. For a somewhat extended discussion of NASTS, upon which these comments are drawn and based, see my "National Association for Science, Technology, and Society," in *Science/Technology/Society as Reform in Science Education*, ed. Robert E. Yager (Albany: SUNY Press, 1996), 291–97. The complete volume provides an excellent summary of STS education, especially at the K–12 level, not only for the United States but around the world as well.

33. For the range of Roy's ideas, see "STS: The Megatrend in Education," *Proceedings of the 1984 International Congress on Technology and Technology Exchange* (Pittsburgh, Pa.: International Congress on Technology and Technology Exchange, 1994); "The Relationship of Technology to Science and the Teaching of Technology," *Journal of Technology Education* 1 (1990): 5–18; "STS: Unsung Revolution in Education," *NASTS News* 4 (1991): 6–7; and "Present Efforts in Technological Literacy," in *Technology Literacy Workshop Proceedings*, ed. Russel Jones (Newark: University of Delaware, 1991), 23–37.

34. Steven Turner and Karen Sullenger, "Kuhn in the Classroom, Lakatos in the Lab: Science Educators Confront the Nature-of-Science Debate," *Science, Technology, & Human Values* 24 (Winter 1999): 5–30.

35. See, for example, Morris H. Shamos, "STS: A Time for Caution," in *The Science, Technology, Society Movement*, ed. Robert E. Yager (Washington, D.C.: NSTA, 1993) and *The Myth of Scientific Literacy* (New Brunswick, N.J.: Rutgers University Press, 1995).

36. On constructivist learning, see, for example, Dennis Cheek, *Thinking Constructively about Science, Technology and Society Education* (Albany: SUNY Press, 1992); Barbara Reeves and Cheryl Ney, "Positivist and Constructivist Understandings about Science and Their Implications for STS Teaching and Learning," *Bulletin of Science, Technology & Society* 12 (1992): 195–99; and Wolff Michael Roth and Michelle K. McGinn, "Knowing, Researching, and Reporting Science Education: Lessons from Science and Technology Studies," *Journal of Research in Science Teaching* 35 (1998): 213–35.

37. Turner and Sullenger, "Nature-of-Science Debate," 21–22, 25. The authors also somewhat chide science studies scholars for not fully appreciating the educational implications of their otherwise "rarified" theorizing.

38. National Science Teachers Association, *Science-Technology-Society: Science Education for the 1980s* (Washington, D.C.: NSTA, 1982) and *Science/Technology/Society: A New Effort for Providing Appropriate Science for All* (Washington, D.C.: NSTA, 1990). Irma Jarcho notes that it was in 1982 that she

helped to found the *Teachers' Clearinghouse for Science and Society Education Newsletter*, which continues today as one of the leading STS curricular publications at the K–12 level. Irma Jarcho, "Thirty Years of STS: Reminiscences of a Survivor," address delivered March 6, 1999, at the annual NASTS meeting, Baltimore, Md., published in *Teachers' Clearinghouse for Science and Society Education Newsletter* 18 (Spring 1999): 1, 17–18.

39. Robert E. Yager, "The Case for STS as Reform," *NSTA Reports!* (May 1991): 9–11.

40. Dennis Cheek, *Thinking Constructively about Science, Technology and Society Education*, 26. Also useful is his unpublished paper, "Education about the History of Technology in K–12 Schools," presented to the Society for the History of Technology Annual Meeting, Pasadena, Calif., October 17, 1997.

41. Bill G. Aldridge, "Improve Science Using 'Basic Science' with Applications," *NSTA Reports!* (May 1991): 8, 10.

42. "Update on STS vs. 'Basic Science' Debate: STS Ahead, but Many Readers Favor Combining Approaches," *NSTA Reports!* (September 1991): 11, 46–57.

43. American Association for the Advancement of Science, *Science for All Americans: Summary—Project 2061* (Washington, D.C.: AAAS, 1989); AAAS, *Science for All Americans—A Project 2061 Report on Literacy Goals in Science, Mathematics, and Technology* (Washington, D.C.: AAAS, 1989); AAAS, *Benchmarks for Science Literacy* (Washington, D.C.: AAAS, 1993); James F. Rutherford, "What's in a Name?" *2061 Today* 1 (1991): 5. Project 2061 has developed, through additional funding from a variety of sources including the NSF, Carnegie Corporation, Mellon Foundation, the MacArthur Foundation, and the Pew Charitable Trusts, a wide range of curricular resources, including a CD-ROM, *Resources for Science Literacy: Professional Development*, and *Designs for Science Literacy,* a guide for systematically approaching K–12 curriculum plans. Information on Project 2061 can be found on its web site at www.aaas.org/project2061 or summarized in its newsletter *2061 Today*.

44. Fred Splittgerber, "Science-Technology-Society Themes in Social Studies: Historical Perspectives," *Theory into Practice* 30 (1991): 242–50.

45. P. Heath et al., "Teaching about Science, Technology and Society in Social Studies: Education for Citizenship in the 21st Century," *Social Education* 54 (1990): 189–93.

46. Cheek, *Thinking Constructively*; Cheek, "Education about the History of Technology in K–12 Schools"; International Technology Education Association, *Technology—A National Imperative* (Reston, Va.: ITEA, 1988); and Mark Sanders, "From the Editor," *Journal of Technology Education* 1 (1989): 3–6; ITEA, *Technology for All Americans: A Rationale and Structure for the Study of Technology* (Reston, Va.: ITEA, 1996); ITEA, *Standards for Technology Education* (Reston, Va.: ITEA, 1998); Kendall N. Starkweather, "The International Technology Education Association (ITEA): A Prominent Voice for Technology Education," *The Journal of Technology Education* 24 (Summer/Fall 1998): 44–47.

47. Thomas T. Liao, "Technology Literacy: Beyond Mathematics, Science, and Technology (MST) Integration," *The Journal of Technology Education* 24 (Summer/Fall 1998): 52–54; New York State Education Department, *Learning Standards for Mathematics, Science, and Technology* (Albany: New York State Education Department, 1996).

48. On this last point, for example, see Langdon Winner, "Conflicting Interests in Science and Technology Studies: Some Personal Reflections," *Technology in Society* 11 (1989): 433–38.

49. For an interesting example of the Rashomon Effect being used to frame the analysis of a controversial STS issue, see Allan Mazur, *A Hazardous Inquiry: The Rashomon Effect at Love Canal* (Cambridge: Harvard University Press, 1998).

50. David Edge, "Reinventing the Wheel," in *Handbook of Science and Technology Studies*, ed. Jasanoff et al., 12.

5

Why Do STS, or Where Do We Go from Here?

> *[W]e must teach [our children] to live in technology and at the same time against technology . . . to develop a critical awareness of the modern world.*
>
> — Jacques Ellul, *Perspectives on Our Age*

> *[STS] occupies an enormous void, one created by a society that long ago committed itself to forge ahead full bore with scientific and technological advance, but never to forge ahead in developing the critical self-reflection such change seems to require.*
>
> — Langdon Winner, *The Whale and the Reactor*

After the question, "What is Science, Technology, and Society all about anyhow?" the second most asked question is, "Why should I study STS?" or its corollary, "What can I do with STS?" The preceding chapters should have provided some inkling of why at least some STS scholars and activists do what they do. It is in this vein that Langdon Winner suggests STS can fill "an enormous void," one from which "critical self-reflection" regarding society's technoscientific goals has largely been missing. This chapter will address the issue a little more directly and from a somewhat more personalized perspective than has characterized the previous discussion, drawing frequently upon examples related to environmental issues.

Framing a Response

I have argued here and elsewhere that, while STS today involves a number of "subcultures," the field received its primary impetus from the widespread social upheavals that occurred during the late 1960s and early 1970s.[1] At the same time, STS, or any other field, including any given aspect of "pure" scientific research, is worth studying in its own right. I therefore have no qualms about "academic" or, if you prefer, "High Church" scholars who might wish to pursue "esoteric" research on a highly rarified scientific case study with no obvious or immediate payoff in terms of policy issues or normative guidelines for society. Nonetheless, my own predilection is for studies that have at least some application, whether it be in the realm of education, policy, or public activism regarding science and technology. As a historian of technology, I appreciate what that field can tell contemporary society about how we got where we are today technologically, including the societally constructed, contingent, and contextualized nature of technology. It may also suggest questions regarding technology that we might otherwise forget to ask without historical perspective. Similarly, STS as an interdisciplinary endeavor offers deep and potentially valuable insights into the nature, applications, and implications for society of both science and technology, if only we are wise enough to incorporate them while constructing a better world.

Numerous scholars have spoken to this need at one point or another in their work. For example, Jesse Tatum, a politically oriented energy studies scholar, has argued that practical concerns, while not predominating in all STS scholarship, do constitute what he calls the "core" of STS as a "social movement," a term he borrows from sociologist Susan Cozzens.[2] Both Tatum and Cozzens argue that STS is a response to what is perceived as the "problematic" nature of science and technology in contemporary society. All those who respond to this "STS Problem"—whether they be STS academics, policy makers, industrialists, or public interest activist groups—are "involved in both thought and action in relation to science and technology in society," although the balance between the two emphases varies depending upon one's position. Most STS academics, according to Cozzens, tend to be "thought-specialists," people who, she believes, would do better if they would interact more directly with those who are more activist in their orientation, so that their intellectual work does not become "isolated, fragmented, and [hence] weak."[3] Whereas Cozzens sees a valuable growing body of applicable interconnected knowledge, which she calls "STS Thought," emanating from STS scholars, Tatum expresses a concern that those "science studies academics" or "constructivists," as he uses the term, may, through their inattention to issues of power and social consequence, in fact, be an "outright distraction" to "core" scholars. Philosopher Steven Goldman has also echoed a similar refrain. He notes that, once well established, STS, somewhat ironically, has turned "self-referentially" inward—away from its original "critical" study of things and praxis, preferring instead to develop and examine theoretical models of technoscience.[4] Leaving aside for the moment the applicability of STS scholarship, collectively Tatum, very explicitly, and Cozzens and Goldman, somewhat less so, emphasize the commitment to democracy inherent in STS as a social movement.

Even such a "High Church" scholar as the term's coiner, Steve Fuller, conceives of STS as a "movement," one that "produc[es] knowledge that enables (and disables) certain transformations of social life." In his view, what is needed is an academically based movement, one based upon "a continuous tradition of social criticism." Here he means to suggest that the "sociological revolution" that has come to dominate much of S&TS scholarship can (indeed, in some sense it already has) provide us with the *"ethnographic distance"* to understand science and technology. "Our task now is to insinuate that swirl of discourse into the pressing social problems of our times."[5]

It is in a somewhat similar vein that sociologist David Edge, editor of *Social Studies of Science,* one of the leading scholarly academic journals, argues for the relevance of the "vision" and "original aims of the STS pioneers." Following the growth in, and sophistication of, STS scholarship, he suggests that, "Perhaps the next phase in the development of STS must be a more urgent concern for *communication* and *translation*: for 'making real' its true potential." He advocates drawing on the complete range of STS scholarship, including the "critical" or "self-awareness" approach that views "science and technology as essentially and irredeemably *human* (and hence social) enterprises," and seeking "reconciliation" with the more "technocratic" wing of STS, to offer prescriptive advice within the political and educational arenas.[6]

Despite its potential for contributing to the democratic process, many of the more activist-oriented STS people, including even Fuller, fear that as STS becomes ever more "professionalized" in an academic sense, it may lose this very "critical" faculty, even becoming coopted by the technoscientific establishment and practices that it seeks to understand and to critique. David Noble, the author of several works highly critical of that establishment, has noted precisely this phenomenon. He suggests that "STS has served to moderate and coopt critical debate about science and technology."[7] In somewhat similar vein, Langdon Winner has noted that his experience in the Department of Science and Technology Studies at Rensselaer Polytechnic Institute has been quite different than at previous institutions with which he has been associated. While RPI's STS department consists of a large number of supportive and largely like-minded scholars, at least as to the critical value of STS, at times it feels as though they are far more isolated, speaking mostly to themselves, rather than about issues of importance to the public at large.[8] Yet, it is precisely the RPI model that Fuller suggests as most conducive to "fostering [the] dynamic sense of credibility" necessary to provide a sufficiently stable platform based on STS scholarship. He believes such "STS Thought," to return to Cozzen's term, reveals that "science and technology have no *intrinsic* powers, but require the active engagement of many people and things, most of which are left out of the standard technoscientific stories, but which, if included, would highlight the politics of the situation."[9]

Cozzens and Fuller have argued that STS creates a useful body of "STS Thought," a body of knowledge, largely sociological in nature, entailing "a continuous tradition of social criticism."[10] If, as I have argued, some sort of contextual or social construction interpretation of science and of technology has become a, if not the, central analytical framework, then what lessons or applications can we make? How to apply that constructivist knowledge in socially redeeming ways is not immediately self-evident, and is something with which many social constructivist scholars have struggled, if not largely avoided. Winner, in a thoughtful, hard-hitting, and subsequently controversial essay, has suggested that this body of STS research dedicated to opening up the so-called black box of technoscience, while valuable in many ways, has turned out to be, if not completely empty, nonetheless, remarkably "hollow." He is quick to admit that social constructivism is an "important school of thought" and that to ignore its findings would, in effect, be foolish. Especially important is the now considerably detailed and nuanced understanding of technoscience as a nonlinear, "multicentered, complex process," in which "choices are available." Thus, "technological development is not foreordained by outside forces but is, instead, a product of complex social interactions."

That said, Winner is just as quick to critique social constructivism, and its largely descriptive orientation, for its failure to enhance sufficiently our "grasp of human experience in a technological society," or at least its agnosticism with regard to "technology and human well-being." For him, "[t]he most obvious lack . . . is an almost total disregard for the social consequences of technical choice." He also laments the constructivists' failure to include all relevant social groups, for example workers, their failure to address structural and cultural dynamics such as issues of class and race, and finally their failure to offer "anything resembling an evaluative stance or any particular moral or political principles that might help people judge the possibilities that technologies present."[11]

Whereas Fuller views the otherwise "admirable" sources, represented by the likes of activist-oriented STS scholars such as Mumford, Ellul, and even Winner himself, as somewhat ill-suited for "sustaining a social movement" because their works "were written pretty independently of one another," and hence fail to provide the necessary "continuous" body of critical thought, it is precisely to these sorts of individuals that Winner suggests turning for guidance.[12] Here I would suggest that, at least in one significant way, Fuller has perhaps overlooked the continuous and "collective knowledge production" that has occurred among such scholars. For example, Ellul referred at a number of points to the work of Mumford, just as Winner himself assessed and extended the ideas of Ellul in his nuanced and scholarly discussion of technological autonomy.[13] Likewise there has been a continuous tradition, especially in the philosophy of technology, that has built upon such scholarship, in the way Fuller believes is necessary for STS to succeed as a social movement. Whether he is unaware of this body of scholarship or has chosen to ignore its potential is unclear, but he is perhaps not alone in this regard, for a quick glance at the extensive bibliography of the 4S-sponsored *Handbook of Science and Technology Studies* reveals few references to works in the philosophy of science and technology. Such lacunae reflect the S&TS origin and compilation of the *Handbook*, but it may also suggest the constrictive disciplinary bounds within which much of STS still labors.[14] All this is not to point fingers, except by way of suggesting that there is much to be learned by transcending those boundaries, which is certainly the case if we hope to develop the vigorous "creative tension" called for by David Edge. Here I think is a reason for the continued interdisciplinary study of STS, because it is pretty clear that we have not learned all there is to be learned from each other, let alone succeeded as Fuller hopes in "insinuat[ing] that swirl of discourse into the pressing problems of our times."[15]

The problem as I see it is not that we have had too many constructivist studies, nor that there is not more to be learned from that avenue of research, but rather that we need to move beyond a level of analysis that is largely descriptive in orientation, to one that is more prescriptive.[16] Thus,

I tend to agree with Winner when he responded to a fairly sharply worded critique of his essay by Mark Elam who accused the former both of having a "blind aversion to social constructivism" and of seeking "to liberate us by converging on the truth." Winner is no more "anticonstructivism" than he is "antitechnology," nor does he seek "to impose universal standards of judgement." As he argued in *The Whale and the Reactor*, that is someone else's argument, or as he puts it in his reply to Elam, the "wrong decal." Instead, the issue is "how to expand the social and political spaces where ordinary citizens can play a role in making choices early on about technologies that will affect them. If we take notice of these matters only after the bulldozers arrive or fiber optic cables are strung, it is simply too late, too late for anything, except for Ph.D. dissertations, journal articles, or book series on social construction."[17]

How to involve the public, which in fact consists of many different "publics," in such matters is of course a complex matter, one that bespeaks of no simple answer, and hence no single solution. What is needed is a range of participatory responses suited to the context at hand. Nonetheless, at root of most, if not all, STS thinking in this regard is some level of commitment to democratic participation.

In the rest of this chapter I would like to turn attention to a few examples suggestive of what STS as an academic field of endeavor and as a social movement might accomplish in this regard. No one of these approaches by itself is likely to be sufficient to the task at hand, nor universally applicable, but taken together they do suggest alternative avenues of pursuit, avenues that assume the use of a set of constructivist "wheels," rather than the positivist turn of mind that admittedly continues to drive much of society's view of technoscience, even in the face of the best STS scholarship.[18] In suggesting some possible routes, I will loosely follow my earlier tripartite view of the field to indicate ways in which each of the subcultures might have a role to play. How the "tension" among the constituent parts of STS can be "resolved," in the sense of being "creatively" stabilized, seems central to the value of STS study and why it is worth pursuing.

Crossing over the High Church–Low Church Aisle

For STS, especially as it has developed within the academy, to have much societal consequence, it is necessary, and even fruitful, to begin within the "academic" corner of STS where most of the so-called constructivist case studies reside. If we can accept, at least for the sake of argument, that these studies have in the main enhanced our understanding of technoscience as an inherently value-laden, multifaceted, and complex process, which suggests the real possibility of societally shaping science and technology, the question remains how best to move beyond the warehousing of ever more sophisticated cases. To translate effectively this already large, accumulated body of STS Knowledge, it is possible for those STS academics "critical" of the technoscientific society, as Mitcham would identify them, to push outward from their scholarship by outlining normative guidelines for action.[19] Several examples of recent scholarship that I find instructive are illustrative of this movement.

All science and technology is political, at least in the broad sense of reflecting power relationships, as well as in the narrower sense of policy and politics. This is what Langdon Winner had in mind when he answered in the affirmative his own question, "Do Artifacts Have Politics?"[20] Recently Daniel Sarewitz, an earth scientist and very much a "realist," in a book entitled *Frontiers of Illusion: Science, Technology, and the Politics of Progress*, has extended Winner's acute observations. He asks how, in a post–cold war environment, "[scientific] research can best serve society" in ways that advance the interests of the public as a whole, not merely those of science professionals.[21] To do so he examines in detail five "myths" that have guided scientific advance for the past fifty years, while suggesting a "new mythology" to guide the scientific enterprise in the next half century.

The myths are as follows:

1. *The myth of infinite benefit:* More science and more technology will lead to more public good.
2. *The myth of unfettered research:* Any scientifically reasonable line of research into fundamental natural processes is as likely to yield social benefit as any other.
3. *The myth of accountability:* Peer review, reproducibility of results, and other controls on the quality of scientific research embody the principal ethical responsibilities of the research system.
4. *The myth of authoritativeness:* Scientific information provides an objective basis for resolving political disputes.
5. *The myth of the endless frontier:* New knowledge generated at the frontiers of science is autonomous from its moral and practical consequences in society.[22]

The thrust of Sarewitz's critique is that these positivist myths have been self-serving of the interests of the scientific community "but often fail to serve the interests of society as a whole." Taking a cue from the constructivists, he notes that we cannot divorce what goes on inside the laboratory from the broader societal context in which it is deeply embedded. "[R]esearch and utilization are in fact inseparable," but not in the simplistic linear sense of more science equals societal progress. Instead there is a need to create both "a more realistic level of expectations" regarding the societal promises made on behalf of the R&D system, and "an increased capacity to achieve societal goals." Critiquing the limitations of science as a "surrogate" for thoughtful social and political action, Sarewitz turns to his own suggestions for an alternative mythology to better serve the public interest. They "focus on the creation of a more explicitly permeable boundary between the laboratory and the surrounding world" and recognize that "the R&D system is an integrated component of the world that surrounds it" and include the following:

1. redoubled efforts to increase diversity within the R&D community—and especially its leadership;
2. fuller integration of the human element in the sense of directing, responding to, and controlling growth and productivity, not merely producing more;
3. development of "honest brokers"—perhaps small, independent, satellite policy institutes—to help create and sustain better harmony and fuller integration of information and expectations between the laboratory and policy arena;
4. creation of enhanced democratic avenues for public participation in science and technology decision making; and
5. an enhanced global focus within the R&D community, one that focuses on "sustainability" rather than on endless economic growth.[23]

Even taken together, these suggestions will not easily, and certainly not immediately, change the U.S. R&D system, let alone the broader authoritarian and restrictive view so characteristic of our technoscientific world view today. To do that will take not only the publication, but also the thoughtful reading and absorption of many other insightful works, that will offer otherwise well intentioned people the needed frameworks for rethinking the place of technology in a human-centered world.

To give a more philosophically oriented example of how academic scholars can contribute, I think David Strong's *Crazy Mountains: Learning from Wilderness to Weigh Technology* is particularly insightful in terms of advocating "vision." Strong's *Crazy Mountains* is a philosophically informed analysis of technology written from the perspective of what wilderness can teach contemporary society about right living. At its most basic, Strong's book is a plea to preserve the Crazy Mountains in south central Montana, although he also argues for the importance of wilderness preservation generally. Beyond this, however, the book is a challenge more broadly "to rethink, to weigh, our vision of technological culture" using wilderness as a guide. To do so, however, requires an understanding of the very nature of technology, and herein lies perhaps the most provocative and valuable of Strong's levels of analysis. In constructing a theory about technology and its consequences for both the environment and our lives, Strong draws on a body of scholarly philosophical literature including the work of Martin Heidegger and especially Albert Borgmann, as well as the work of less academic writers such as the ecologically minded wildlife manager Aldo Leopold, suggesting that, in fact there has been a "continuous" tradition of a prescriptive social criticism, something that Fuller had suggested as missing from much Low Church writing.[24]

In seeking to find ways "to build a culture set in the context of technology yet ordered by things as opposed to commodities," Strong acknowledges his intellectual debt to Borgmann. The latter suggests a distinction between "focal things" (e.g., a fireplace) and technological "devices" (e.g., a central heating thermostat). The first "is inseparable from its context," while the latter tends to "split" ends and means. Strong views technology, at least in terms of the collective totality of "devices," as being incompatible with environmental and human well-being. He seeks instead what he calls "correlation coexistence," a reciprocal connection between human beings and things, which will allow our culture to move beyond its current "vision of the good life [as] the goods life." For Strong, the antidote to society's heedless attraction to "the technological good life" is the "tonic" of wilderness; without the latter, we cannot be "at home" in the world. The answer is not to do away with technology, which would at once be impossible and undesirable, but rather to "rebuild" in ways that are "appropriate and vital." Starting at the individual and community levels, we must "emancipate ourselves" from "the pointless character of consumption as a way of life," for then "things . . . not technology [will] order the world."[25]

Strong's book is not only thoughtful in its analysis of technology, but also thought provoking. This is due, at least in part, to Strong's own "inspirational" nature writing, which is some of the best prose in the book. Thus, independent of whether one accepts fully his assessment of technology's place within society, I believe Strong's effort does suggest the possibility of drawing on "models of collective knowledge production," to borrow Fuller's phrase, but of also transcending it to offer prescriptive suggestions regarding society's pressing problems.

Anthropologist David Hess has shown the applicability of STS academic knowledge in yet another way through his "intervention-oriented" research on an alternative bacterial cancer theory. The relative failure of the last generation's "war on cancer" suggests the need to make hard decisions about future expenditures and treatment policies. Hess believes "outside" or "third-party" social science involvement can at once provide assistance in the direct evaluation of the scientific merits of different positions on a controversial issue and offer an alternative successor framework of analysis to SSK.

Four principles of analysis, which Hess expects will be applicable in other situations, evolve out of the cancer study. First, the analysis is "political" in its exploration of "the operation of power in the history of a field of knowledge," that is, in this case, how one cancer theory became suppressed within the broader body of cancer research. Secondly the analysis is "cultural" in that it develops a "noninstrumentalist explanation . . . of the dynamics of power" described in the first step. Here the emphasis is on "the role of evidence and efficacy in shaping the field of possibilities." Thus, while science is viewed as a rational activity, sociocultural factors, such the role of gender and transcultural movements, nonetheless shape preferences and contribute to the construction of consensus. Third, the analysis is "evaluative" in that "it draws on the philosophy of science to weigh the accuracy, consistency, pragmatic value, and potential social biases of the knowledge claims of the consensus and alternative research traditions." This prescriptive approach is neither relativist nor realist, but rather assumes "evidence can be established but always within a social situation," which allows for reinterpretation. Finally, the analysis is consciously and openly "positioned" through the evaluation of various agendas for action, in this case institutional changes and future research investments, which offers social scientific perspective, without necessarily allying oneself completely with a specific position. I view this as another excellent example of how STS Knowledge can transcend the seeming academic/activist schism.[26]

There are of course many other studies that transcend this division. Winner himself pointed to the work of historian Ruth Cowan, who has shown how choices regarding household technologies have consequently affected social life, and that of mathematician turned social scientist Brian Martin, who has suggested the need to write in "more accessible fashion" and to provide "practical materials that can be taken up by practitioners or activists." In this vein a number of STS scholars have begun to call for "constructive technology assessment." By this they mean a shift from technology assessment that goes beyond earlier attempts to forecast the potential impacts of technology, both positive and negative, to a form of assessment, or active management, that proactively opens up the technoscientific planning and decision-making process to include a broader range of people and perspectives, that recognizes the inherent uncertainties, and that leaves open to as great a degree as possible alternative courses of action should subsequent knowledge warrant change. One could also point to Eduardo Aibar's application of a constructivist framework to study Barcelona's nineteenth-century Extension to the older walled city center. Written in part as a response to Winner's concerns, including an analysis of the working class as a relevant social group, the study draws on Weibe Bijker's notion of "technological frames," and "socio-technical ensembles" to suggest a way of rethinking concepts like "power" and analyzing "the co-production of technology and society." Somewhat similarly, sociologist Steven Yearly, in an essay in the 4S *Handbook* surveying much of the existing science studies literature on the environment, concluded that "science studies provides a framework for explaining the ambiguous role of science in illuminating problems or resolving conflicts about the environment . . . and that the study of environmentalism is equally beneficial to science studies . . . not only because the science of the environment occupies an increasing percentage of scientists' attention but because features of environmentalism throw issues about science and scientific knowledge into particularly sharp relief." In this manner, it is thereby possible for otherwise "academic" studies to transcend the intellectual median strip separating the explanatory from the more critical or prescriptive side of STS.[27]

Some might argue that studies like those of Sarewitz and Strong, or even Hess, have already veered over into the "Low Church" side of the STS academic aisle, and perhaps they do, but if so, it suggests as much about the inadequacies of such a bifurcation of the field, and points instead to the necessity for what David Edge almost wistfully imagined—a more inclusive, even if not homogeneous, "broad church," one that could draw on a collective body of STS knowledge that includes both constructivist case studies as well as more prescriptive calls for action.[38]

STS Activism

Within the more activist-oriented part of STS are those social interest action groups Carl Mitcham categorized as critical of modern technoscience. Among their wide-ranging concerns are nuclear weapons and energy, consumer safety, and environmental well-being. They include international, national, and regional and local organizations ranging from Greenpeace and the Union of Concerned Scientists to the Sierra Club and national "green" parties to NIMBY (Not In My Back Yard) grassroots groups. At whatever their level, each of them reflects a concern regarding the present, or projected, state of affairs regarding some aspect of science or technology. Each is committed to contributing directly to the political decision-making process regarding the issue(s) at hand, and to that end believes in and promotes enhanced democratic participation to at least some degree. In the context of this discussion of STS, I am not going to delve into such groups, for that is beyond the scope of what I intend and would entail a book unto itself. In addition, most such groups often tend to be reactive, that is, responding to plans or decisions some time after they have been publicly announced, rather than early on in the process before a certain amount of momentum has set in. Instead, I will present two examples of ways by which STS, as an activist-oriented academic field, can contribute to enhanced democratic participation as an inherent part of the decision-making process, not merely as a public relations add-on or in the context of reactionary political controversy.

I want to begin this section by focusing on a somewhat underappreciated book, *Social Responsibility in Science, Technology, and Medicine,* by philosopher of technology Paul Durbin. It should be noted that Durbin is one of the few STS scholars who consistently includes medicine as an area equally as germane as science and technology in his discussions of technoscientific issues. He draws on a rich philosophical literature, especially the ideas of activist philosophers George Herbert Mead and John Dewey who sought the pragmatic application of philosophy within social and political criticism, in a call for more socially responsible, public interest activism, on the part of scientific and technical professionals.

Durbin believes such professionals, including academic philosophers, must "go beyond what is demanded by their professions and get involved with activist groups seeking to bring about fundamental change," if we are to have any hope of resolving contemporary society's technoscientific problems. He believes that technology can be controlled by democratic means, "if enough people have the spirit, energy, and drive to mobilize against technosocial ills." To do so, however, the focus of thought must be local, not global, and will require "community action," action which will also create new meaning for society.[29]

Among the sociotechnical problem areas he tackles are three broad-brush social issues—science education, health care, the media and politics—and four more specific technoscientific ones—biotechnology, computers, nuclear weapons and power, and the environment. Following the social work tradition, in which social workers routinely become engaged in activist causes and movements, Durbin offers both the philosophic rationale and specific mechanisms for activist involvement on the part of technical professionals. By way of example of the kind of philosophic and democratic bridge building Durbin has in mind, his suggestions regarding biotechnology are instructive. Here he calls for the involvement of philosophers, not as "moral experts" per se, but in the role of skilled issue-clarifiers, working closely and democratically with scientists, bioengineers, and biomedical researchers, government regulators, industry representatives, and the lay public. Issues requiring equitable and evenhanded solutions might include assuring that genetic testing not be used in discriminatory ways; working with advocates of biotechnology and EPA regulators to achieve a fair balance between development and ecology; wrestling with the issues of transferring biotechnology to developing nations; developing balanced regulations and evenhanded enforcement satisfactory to both genetics researchers and regulators to name but four possibilities.

Durbin's "meliorist" approach does not entail an unqualified optimism that all problems will be fully "solved," especially in the face of not necessarily being able to change the underlying social causes. Nonetheless, he does see the "possibility" of at least some progress, which opens up the further "possibility that newly empowered citizens will move on from past limited successes to attack head-on the overarching issue of technoeconomic inequities." Like Noble, he warns that we must be careful that such reform efforts not be distorted or subsumed by corporate or political powers and hence merely reinforce the status quo. Especially crucial here is the issue of who gets to set the agenda in all such public participation situations. For example, some industries, especially public utilities because of the nature of their mandates, make active use of citizen advisory panels, a progressive participatory idea with which I am in full accord; however, it is still important not to let this become a mere public relations ploy.[30] In effect, Durbin ultimately is arguing against "the technocratic state," by which he means the mainstream view of modern life into which most Americans and at least a majority of citizens in the developed nations have readily accepted. He believes instead that we should seek to include an ever-widening circle of viewpoints keyed to "justice" within a "flourishing" democratic society.[31]

The individual who has probably done the most from an STS perspective to argue for enhanced democratic participation in the technoscience decision-making process has been Richard Sclove, founder of the Loka Institute, a nonprofit, citizen-action think tank and network. Underlying Sclove's work and that of others who promote enhanced public participation in the science and technology process is a commitment to "strong democracy." The ideas central to this notion are drawn from the work of Benjamin Barber and expanded upon by Sclove in his book, *Democracy and Technology*, in terms of "design criteria for democratic technologies."[32]

The first two and most general criteria set the tone and framework for those that follow. Thus, Criterion A states: "Seek a balance among communitarian/cooperative, individualized and transcommunity technologies. Avoid technologies that establish authoritarian social relations," while Criterion B says: "Seek a diverse array of flexibly schedulable, self-actualizing technological practices. Avoid meaningless, debilitating, or otherwise autonomy-impairing technological practices." Subsequent criteria dealing with democratic politics and self-governance stress the local and the sustainable over the global and exploitive. "[O]rganizing society along relatively egalitarian and participatory lines," Sclove argues, would entail adopting most, if not all, of a series of "strategies" that would include the need to "map local needs and resources," while "reach[ing] out to political movements [to] build coalitions." The initiation of "democratic R&D and design," combined with expanded "civic technological empowerment," would help to "democratize corporations, bureaucracy, and the state." Sclove concludes his analysis by asking a penultimate question: "is it realistic to envision a democratic politics of technology?" Throughout he draws on the Amish by way of a small-scale example, suggesting the answer is "yes," but the more telling point is his final question: "Isn't it unrealistic not to?"[33]

As illustrative examples of his approach to democratizing science and technology, Sclove likes to point to two possible approaches beyond the admittedly limited and religiously motivated Amish. One, which has been in place for some time, is the so-called science shop, found at its most developed state in the Netherlands, the other being what are known as "consensus panels," European-style citizen advisory panels for science and technology policy. Although differing in approach, in both cases the intent is to provide expanded knowledge to, and to allow greater participation by, the general public.

In the case of science shops, which are, in effect, university-based community research centers, academic faculty, staff, and students are available to provide research for organizations, whether they be environmental, labor, or other nonprofit types, which do not have the expertise nor the resources to conduct their own research on issues of local or regional import. Subsequently, such groups make use of this "academic" research as part of their input into the decision-making process, thereby providing a way around the argument that the "public" is not expert enough to contribute knowingly to the deliberations. Presently there are almost forty such science shops in the Netherlands, while numerous other nations including Denmark, Germany, England, and even the United States have developed similar community research centers. Most recently the Canadian Social Science and Humanities Research Council has initiated a national network of twenty-two community-based research centers called CURA (Community-University Research Alliances). There is even a newsletter coordinating activities among the informal network of such centers.[34]

Consensus conferences, pioneered in Denmark and conducted under the auspices of their Board of Technology, offer an opportunity for panels of everyday citizens who are nonstakeholders to inform themselves deeply on given topics in science and technology and then, following open discussion and debate, to reach a decision that is announced publicly as an advisory report at press conferences. Such reports are not binding, but they do stimulate broad popular debate and increase public understanding, and can help change policy and thus acceptance levels. They are offered as advisory input which the Danish Parliament can then act upon as it sees fit. The first such Danish consensus conference was held in 1987, and since then numerous others have been successfully conducted. By way of specific example, a 1989 citizens' panel on the Human Genome Project encouraged support for basic genetics research, but it also called for further work on the societal consequences and influenced the Danish Parliament to enact legislation prohibiting employment and insurance decisions based on genetic information. In March 1999 a Danish citizens' panel examined the issue of genetically engineered foods. While the panel stopped short of calling for a moratorium, they did call for stricter regulatory control, including better consumer labeling practices and restrictions on corporate monopolies with regard to genetic technologies.

At least a dozen nations have now organized or are about to hold such citizens' panels. For example, Japan held a consensus conference on human gene therapy in March 1998 and is planning a second on the "High Information Society." Canada held a conference on food biotechnology in March 1999, while England held its second such meeting on the topic of radioactive waste disposal in May 1999. Other nations such as Australia and South Korea are considering holding similar consensus conferences on topics of import to them, further testifying to the value of this sort of mechanism for enhancing citizen participation in deliberations regarding important and potentially controversial technoscientific issues.[35]

In April 1997 Sclove organized the first such citizens' advisory panel in the United States on an experimental basis with NSF funding and the support of the Massachusetts Foundation for the Humanities and MIT's *Technology Review* magazine among others. Held on the campus of MIT, the conference explored the issue of "Telecommunications and the Future of Democracy." The panel consisted of fifteen citizens reflecting a broad cross section of the greater Boston area, including five people of color. They initially spent two weekends discussing background readings and briefings on telecommunications issues and then listened to ten hours of expert testimony from computer specialists, government officials, and business executives including the president of New England Cable News and the Congressional liaison to the Department of Commerce who helped draft the 1996 Telecommunications Reform Act, as well as public-interest group representatives.

Panel members showed that they were able to assimilate a broad array of written work and expert testimony and integrate this information with their own views to reach a reasoned consensus on such Internet-related issues as policy making, content and standards, universal service, and education and technology. Concerned that the power of the telecommunications industry often abuses the interests of the people, the panel urged "policy makers . . . to anticipate the presently uncharted effects of the new technology, taking into account all aspects of a community. . . . " To assist in so doing, the panel urged the mandatory creation of participatory citizen advisory panels to represent those people who will be affected by decisions. They also expressed concern regarding the protection of First Amendment rights and the right to personal privacy. They advocated "universal service," not just "access"—that is, the ability to take advantage of that access, including the ability to broadcast, as contributing to "the future of democracy." At the same time they noted that such service "means little without education," to which end they called for the widespread installation of computers in the schools, not as an end in itself, but as a means for accessing information, the "foundation of democracy." In the end, even in the face of financial constraints that limited the range of expert testimony to which the panel was exposed, this experiment in participatory citizen deliberation on technoscientific matters proved that the U.S. public is interested in playing an active advisory role and capable of making recommendations useful in practical decision making.[36]

Durbin and Sclove both work in the so-called Low Church, or problem-focused, corner of STS. Although their work is more explicitly activist in its orientation than the majority of STS academics, both draw upon scholarly literature for their intellectual underpinnings. Neither tries to suggest *the* solution or answer in any given situation, for the intent is not to convert anyone to a particular church dogma, but rather to open up the decision-making process to a more democratic range of human-centered perspectives and viewpoints. Thus, the value in the work of Durbin and Sclove lies in their suggestions of possibilities for enhanced democratic participation in technoscientific decision making.

Education for Enhanced Technoscientific Management and Literacy

There are also those within the STS academic community who focus attention either on the improved management of science and technology or on the promotion of an enhanced public understanding of technoscience through educational programs at both the undergraduate and graduate levels. Some programs are designed to turn out more societally thoughtful and ethical scientists, engineers, and business professionals, while others reach out to a much broader group of students, society's future decision makers and voters, through courses taken as general education electives. A few examples drawn from this area of STS, many of which are related to the environment and issues of sustainability, should suffice to indicate the possibilities.

Many science and technology public policy programs focus on the professional training of future technology managers and the education of policy analysts as noted in the previous chapter; however, far fewer devote much attention to societal impacts of issues such as that of sustainable development. Two exceptions to this admitted generalization suggest what is possible, however. The first example is the Center for Energy and Environment at the University of Delaware, which has several faculty who specifically self-identify their interests as including sustainability. The program offers a special course on sustainable development, a subject area that students may pursue as a masters or Ph.D. level research field. A sampling of recent doctoral level research projects includes: "Sustainable Energy Options for the Indian Power Sector" (Chandra Govindarajala); "Toward a Sustainable Energy System for China's Economic Development: The Case of the Electric Sector" (Bo Shen); and "Third World Perspectives on Theories of Sustainable Development" (Subodh Wagle). While not all STPP/SEPP programs are as strongly oriented to energy and environment issues, the example does point to what is possible, and perhaps desirable at least on a limited scale, given the issue's importance to contemporary world politics and public policy. Because many of the graduates of STPP/SEPP programs will subsequently move into positions of importance in government and business, or possibly as analysts of public policy, this seems an especially appropriate opportunity, one that should not be minimized or missed out on.

The second example is somewhat broader and relates to a current effort to draw out the policy implications contained within the dissertations and theses of recent European STS graduates conducted under the auspices of the European Inter-University Association on Society, Science, and Technology (ESST). This research has hitherto gone unanalyzed in any systematic way, and ESST has thus established the POSTI (Policies for Sustainable Technological Innovation in the 21st Century) Project. The project's subtitle—"Lessons from Higher Education in Science, Technology and Society"—indicates the intent to draw out from this "academic" research applications regarding sustainable technological innovation that are relevant to formulating practical future policies. The project is interested both in the sustainability of the innovation process itself, but more broadly in the ways by which the innovation process can contribute directly to improving environmental quality. To that end they have established a workshop to share appropriate research and to discuss sustainability issues. In addition, they have created a database mapping the research and summarizing the main resulting conclusions regarding the policy implications. Again, here is an example of how academic policy-oriented research, thoughtfully applied, can enhance the management and regulation of technological innovation.[37]

To illustrate further why a familiarity with concepts of environmentally sustainable development (ESD) in STPP programs is important, consider two brief specific examples drawn from trips to China in 1992 and 1998.[38] Since the late 1970s and the end of the Cultural Revolution, China has experienced a rapid spread of interest in STS-related studies and education, with an especially strong, applied problem-oriented focus, which is even reflected in the more theoretical aspects of the curriculum.[39] At the time of my first visit, the Chinese government had just determined to proceed with plans to dam the Yangtze River in the scenic Three Gorges area. This decision was the equivalent of earlier, albeit subsequently failed, plans to dam the Colorado River in the Grand Canyon area of Arizona in the United States; it also carried with it the need to relocate several million displaced Chinese citizens. The construction process continues today, as is frequently reported on by the U.S. and world press. My concern at the time had less to do with the decision itself, much as I might personally regret it, as with the seeming lack of open public debate and discussion in the decision-making process, one which in this case reflects real trade-offs between economic development and environmental preservation, if not sustainability per se. Here was a case, even in a nation not known for its democratic decision making, where I believe exposure to ESD issues through STS education for economic planners, developers, and environmentalists alike might have enriched the decision-making process, and the final political outcome, more than I at least sense was the case. At a conference on "Humanistic Factors in the Development of Advanced Science and Technology" hosted by the Chinese Academy of Social Sciences at which I spoke during the more recent trip, I noted that there was far more open discussion of such sensitive science and technology-related issues, at least among the academically oriented people present.

The second example has to do with the Chinese desire for, and apparent move toward, greater adoption of the automobile. At one point during my stay in Beijing in 1992, my hosts from the Institute of Science, Technology, and Society at Tsinghua University, quite proudly, and rightfully so, described a research project in which their graduate students had successfully played a role in helping to plan and lay out the athletic complex and a new peripheral highway built to accommodate the housing and related traffic associated with the 1990 Asian games held in Beijing, the same complex that would have been used had China been successful over Australia in its bid to host the Summer Olympics in the year 2000. While I applauded this work, subsequent discussion and my suggestion, based on U.S. experiences, that expanded urban and interstate highway systems might actually increase road traffic and only lead to more congestion than that which already existed met with a certain amount of incredulity. My admittedly impressionistic observations from the more recent trip appear to confirm that the country has indeed moved ahead with expanded and more individualized, automotive transportation, even though most of the cars are still owned and operated by businesses and official institutions. Unfortunately traffic congestion and transport-related emissions appear to have correspondingly worsened. Thus, one must ask whether this sort of development is environmentally sustainable, not only in terms of pollution, but also in terms of energy use and the related increase of land devoted to roads, highways, and parking facilities, land that in some cases may well be needed for agricultural production and living space. Again, without attempting to pass judgement on the "correct" response, I want to suggest that this is clearly an example of where sustainability and other related STS issues are at stake, and to which well-trained STPP/SEPP scholars and planners could directly contribute.[40]

In addition to the more policy and management-oriented approaches to science and technology just discussed has been an increasing concern for the education of engineering and technically oriented students. Here the operative emphasis is not on "training," but should be on "education" in the broad integrative sense of preparing students by "combining the cultural with the professional, the theoretical with the practical, and the humanistic with the technical in a modern, liberal education that serves as preparation for a useful life."[41] While this extract from Lehigh University's mission statement clearly has wider application, it is in this broader sense of "education" that traditional engineering training, if only because of the sharp constraints of the curriculum, has frequently failed its students, and by extension the society they serve. By failure here I mean that because of the extensive scientific and technical requirements embedded in their education, young engineers are very likely going to be less exposed to societal, including environmental, issues than are their humanities and social science peers. Recognizing this drawback, the Accreditation Board for Engineering and Technology (ABET) has in recent years established guidelines regarding engineering students' required humanities and social science course work that recommends the value of STS type courses.

One of the more popular and valuable manifestations of this focus of concern has been the rapid expansion of an interest in, and a growth in the teaching of, engineering ethics. Several of the past annual meetings of the National Association of Science, Technology, and Society have contained pedagogically oriented sessions devoted to the presentation of ideas and materials related to engineering ethics. One of the largest of such efforts has been the Engineering Coalition of Schools for Excellence in Education and Leadership, a consortial project involving some ten universities across the United States. The project tries to improve engineering education through curricular developments that emphasize the "contextual" nature of engineering for students; to enhance their appreciation for the ethical issues engineers encounter; and to expand the "design" experience beyond the merely technical to include societal issues as well.[42]

Engineers must learn to go beyond merely "designing for manufacturing and assembly" and to adopt a more holistic mentality, sensitive to design for society and the environment, one in which they think in terms of "cradle to grave" and "life cycles."[43] In this regard, I want to mention several interesting developments related to engineering and technical design education that are suggestive of one direction in which STS can go. One is Rensselaer Polytechnic Institute's program in Product Design and Innovation, which is, in effect, a dual major in Engineering Science and STS, which integrates an emphasis in hands-on design with courses that discuss "how a product will integrate into different cultures and societies, and what its ultimate impact might be." A second is the University of Virginia's creation of an interdisciplinary Institute for Sustainable Design, with an analogous effort at Pennsylvania State University, to inculcate a "green design" awareness among its engineering students. Currently this is being done through the holding of an annual Green Design Conference and by requiring all projects pursued within their introductory computer-aided design course to take into account the environmental "footprint" or impacts. Finally, the Colorado School of Mines in its

Mission and Goals Statements explicitly "dedicate[s] itself to responsible stewardship of the earth and its resources," and notes specifically its commitment to "reduc[ing] the world's dependence on non-renewable resources." It actualizes this set of lofty goals by committing itself to educate students who "exhibit ethical behavior and integrity" and will "become responsible citizens . . . , particularly through stewardship of the environment." It does so at least in part through courses such as that entitled "Nature and Human Values," which includes units on sustainability, history of science and technology, and "alternative views."[44]

For any of this environmental concern to have applicable meaning, of course, it must be reinforced beyond the classroom. One place where this should take place, but where little has yet been accomplished is within the professional engineering organizations. One organization that does this is the Institute of Civil Engineers of Spain, whose Environmental Commission has developed a decalogue of environmental commandments. Among them is the injunction to "love and respect nature, particularly the physical environment in which you live, work, and design (develop)" and the injunction "to conserve and improve the natural environment in your professional activities. . . . "[45] Another such organization is the Institution of Engineers, Australia, which has developed a number of policy documents entailing environmental guidelines and principles for engineers, including a commitment to sustainable development. As spelled out in their 1994 Code of Ethics, "Engineers, because of their professional role in society, have a particular obligation towards the integration of development and the environment, leading towards sustainable development."[46] While surely there are other issues that are equally important and require a broadening out of engineering education and an enhanced ethical vision of what it means to be a professional engineer, the ESD example suggests what is both necessary and possible regarding the societal "contextualization" of science and technology.[47]

Finally, without repeating the arguments of the section on general education in the previous chapter, let me reiterate that general education may well be the proper place for most students to be exposed to the myriad ways in which society shapes scientific research and technological development, and also how science and technology affect society. If STS—as either academic field or social movement—is to have any influence in our ability to shape and control technoscientific society, then STSers must insure a broad-based understanding of the interrelations between ideas, machines, and values at the level of general education. In some ways this is already foreordained, for the science-technology-society relationship is too complexly interwoven for the educational process to ignore. Nonetheless, we can and must work even harder to envision, understand, and shape the complexities—and, indeed, the beauties—of STS relationships in working toward the construction of a better world. To be successful in this task, STS must incorporate in a balanced way all three legs or emphases within STS—academic scholarly research into the nature of the science-technology-society web of relationships; socially responsive, technoscientific management and policy; and activist-oriented participation in response to the problematic natures of science and technology. If we fail in this, as suggested at the beginning of the chapter, much of the past generation of both STS scholarship and activism will largely have been for naught.

Notes

1. See chapter 1 of this volume and Stephen H. Cutcliffe, "The Emergence of STS as an Academic Field," *Research in Philosophy and Technology* 9 (1989): 287–301.

2. Jesse Tatum, "Science, Technology, and Society: Issues in Professionalization and the Future," unpublished colloquium paper presented at Rensselaer Polytechnic Institute, February 2, 1995; Susan Cozzens, "The Disappearing Disciplines of STS," *Bulletin of Science, Technology & Society* 10, no. 1 (1990): 1–5, also revised and reprinted in *Visions of STS: Contextualizing Science, Technology, and Society Studies*, ed. Stephen H. Cutcliffe and Carl Mitcham (Albany: SUNY Press, 2000); and Cozzens, "Whose Movement? STS and Social Justice" *Science, Technology & Human Values* 18 (Summer 1993): 275–77.

3. Cozzens, "The Disappearing Disciplines of STS," 2.

4. Goldman bases his concern on an assessment of dramatic shifts over the course of the 1990s in the ways "industry" produces goods and services, what has become know in some circles as "agile manufacturing." In this mode, mass production is largely passé, with most corporations, and hence the technologies involved, focused on serving rapidly changing "niche" markets, the ultimate being markets of "one." Thus, by way of example, Dell *builds* computers only *after* they have been ordered and paid for, and it manufactures *nothing*. Such an approach, which is becoming increasingly widespread, utilizes "constant flow" without any real inventory, in turn requiring short lead times and flexible, but close, linkages between suppliers and assemblers that might be located anywhere around the globe. This shift entails dramatic STS-related implications (sociopolitical, economic, environmental) which are hardly recognized by most inward-looking STS theorists. See Steven L. Goldman, "The Agility Revolution: STS in the New Industrial Order," paper presented March 5, 1999, Annual NASTS Conference, Baltimore, Md. For a fuller explication of the revolution in agile manufacturing, see

Steven L. Goldman, Roger N. Nagel, and Kenneth Preiss, *Agile Corporations and Virtual Organizations: Strategies for Enriching the Customer* (New York: Van Nostrand Reinhold, 1995); and Steven L. Goldman, Kenneth Preiss, and Roger N. Nagel, *Cooperating to Compete: Building Agile Business Relationships* (New York: Van Nostrand Reinhold, 1996). Also useful as a brief introduction to this topic and its relationship to STS is Wilhelm E. Fudpucker, "Postmodern Production and STS Studies: A Revolution Ignored," in *Visions of STS*, ed. Cutcliffe and Mitcham.

5. Steve Fuller, "STS as Social Movement: On the Purpose of Graduate Programs," *Science, Technology & Society Curriculum Newsletter* 91 (September 1992): 15.

6. David Edge, "Reinventing the Wheel," in *Handbook of Science and Technology Studies*, ed. Sheila Jasanoff et al. (Thousand Oaks, Calif.: Sage, 1995), especially 4–5, 15.

7. Personal letter, David Noble to author, September 25, 1998. Noble also generally declines to take membership in those scholarly organizations devoted to the study of science and technology. He notes in a letter to Alex Roland, a past president of the Society for the History of Technology, for example, that "a focus on the history of technology per se ran the risk of being counterproductive in the long run in that it tended to reify and isolate technology and thereby render more difficult a fuller understanding of technological endeavor." Quoted in Alex Roland, "What Hath Kranzberg Wrought? Or, Does the History of Technology Matter?" *Technology and Culture* 38 (July 1997): 699n.6. See, in particular, Noble's books *America by Design: Science, Technology, and the Rise of Corporate Capitalism* (Oxford: Oxford University Press, 1977) and *Forces of Production: A Social History of Industrial Automation* (New York: Knopf, 1984).

8. Comment by Langdon Winner in the discussion period following his presentation at an STS education conference in Valencia, Spain, June 1989. The presentation was subsequently published as "Conflicting Interests in Science and Technology Studies: Some Personal Reflections," *Technology in Society* 11 (1989): 433–38. Winner has also related his experience of serving on an Ethics and Values in Science and Technology (EVIST) review panel for the National Science Foundation in the early 1980s. At that time the panel was told by a senior NSF administrator that a major reason for EVIST's support within the NSF was that, in Winner's summarizing words, "it helped the scientific community respond to political pressures from Congress and the general public." In other words, the public's misgivings could be intentionally assuaged by pointing to various STS teaching and research programs where, again in Winner's words, "all the social and ethical perplexities . . . were being rigorously investigated." "The Gloves Come Off: Shattered Alliances in Science and Technology Studies," in *Science Wars*, ed. Andrew Ross (Durham, N.C.: Duke University Press, 1996): 107–8.

9. Fuller, "STS as Social Movement," 4.

10. Fuller, "STS as Social Movement," 5.

11. Langdon Winner, "Upon Opening the Black Box and Finding It Empty: Social Constructivism and the Philosophy of Technology," *Science, Technology, & Human Values* 18 (Summer 1993): 374–75, quotations 362, 364, 366, 368, 371–72, 375. On the issues of race and class, see also Susan Cozzens, "Whose Movement?"

12. Cozzens, "Whose Movement?" 367, 375; Fuller, "STS as Social Movement," 5.

13. See, for example, Jacques Ellul, *The Technological Society*, trans. by John Wilkinson (New York: Knopf, 1964), passim; and Langdon Winner, *Autonomous Technology: Technics Out-of-Control as a Theme in Political Thought* (Cambridge: MIT Press, 1977).

14. For a more extended review of the *Handbook*, see my generally enthusiastic review, "A Hitchhiker's Guide to STS," *Technology and Culture* 36 (October 1995): 1015–20; and the somewhat more critical comments of Carl Mitcham, Rudi Volti, and Wilhelm Fudpucker regarding the boundaries set for the project, and hence its omissions, in a special review section of the *Science, Technology & Society Curriculum Newsletter* 106 (Winter 1995): 1–11.

15. Here one might argue, of course, that while coherent thought was certainly involved, the successes of such social movements as that of labor, or even the antinuclear movement, did not necessarily require a continuous body of academic critical thought.

16. I do not want to give the impression here that I believe there is nothing further to be gained by descriptive sociological case studies, for I believe there is more such fruitful research that could be done. For example, in his science policy-oriented survey of the STS field, Andrew Webster suggested a number of currently understudied areas worthy of future sociological research, including what he called the "political economy," that is, the working relationships, of the laboratory, especially among the junior scientists and technicians, and the organization and culture of corporate R&D laboratories. *Science, Technology, and Society* (New Brunswick, N.J.: Rutgers University Press, 1991), 157-58.

17. Mark Elam, "Anti Anticonstructivism or Laying the Fears of a Langdon Winner to Rest," *Science, Technology, & Human Values* 19 (Winter 1994): 101–6; and Langdon Winner, "Reply," *Science, Technology, & Human Values* 19 (Winter 1994): 107–9. For Winner's comment on what it means to be "antitechnology," see chapter 1 of this volume, and Winner, *The Whale and the Reactor*, xi.

18. Here I am borrowing David Edge's metaphor of the "remorseless wheel" of the positivist "received view"of science and technology. See his "Reinventing the Wheel," especially 4–5, 18–20.

19. Carl Mitcham, "Science-Technology-Society in Theory and Practice: A Conceptual Introduction," in *Spanish Language Handbook of Science, Technology and Society*, ed. Andoni Alonso et al. (Estella, Spain: Editorial Vervo Divino, 1996), 9–12. Richard Gosden, in his essay "STS Whipping Posts Enclose the Discipline," *Technoscience* 8 (Fall 1995): 14–15, refers to this same basic group or perspective in terms of "scientific relativism." Chapter 4 of this volume contains a somewhat more extended discussion of these positions.

20. Versions of this essay have appeared in several places, including originally in *Daedalus* 109 (Winter 1980): 121–36, but it is most readily accessible as chapter 2 in *The Whale and the Reactor.*

21. Daniel Sarewitz, *Frontiers of Illusion: Science, Technology, and the Politics of Progress* (Philadelphia: Temple University Press, 1996), x. Sarewitz served in Washington, D.C., first as a AAAS Congressional Fellow and then as a science consultant for the House Committee on Science, Space, and Technology. He is currently a senior research scholar in Columbia University's Science, Policy, and Outcomes Project.

22. Sarewitz, *Frontiers of Illusion*, 10–12.

23. See Sarewitz, *Frontiers of Illusion*, chapter 9 for Sarewitz's policy suggestions; quotations from 14, 114, 171–72.

24. David Strong, *Crazy Mountains: Learning from Wilderness to Weigh Technology* (Albany: SUNY Press, 1995), 9.

25. Strong, *Crazy Mountains*, quotations 12–13, 80– 82, 166, 207, 215. For an extended analysis of Strong's book, see my review essay, "The Importance of Being Crazy," *Research in Philosophy and Technology* 16 (1997): 211–16. For Borgmann's ideas, see his *Technology and the Character of Contemporary Life* (Chicago: University of Chicago Press, 1984) and his more recent *Crossing the Postmodern Divide* (Chicago: University of Chicago Press, 1992).

26. David Hess, *Can Bacteria Cause Cancer? Alternative Medicine Confronts Big Science* (New York: New York University Press, 1997). In outlining Hess's analytical framework, I have drawn upon the summary contained in his recent survey, *Science Studies: An Advanced Introduction* (New York: New York University Press, 1997), 152–55. On the issue of the societal embeddedness of the war on cancer, see also Robert Proctor, *Cancer Wars* (Cambridge: Harvard University Press, 1995), in which the author argues we could "cure" 30 to 40 percent of cancers by outlawing cigarettes, and perhaps eliminate another 30 percent through more stringent environmental laws, all at a cost far less than the billions of dollars spent on the "War on Cancer."

27. Winner, "Upon Opening the Black Box," 377n.6. Ruth Schwartz Cowan's *More Work for Mother: The Ironies of Household Technology from the Open Hearth to the Microwave* (New York: Basic Books, 1983) is perhaps her best known work in this regard. Brian Martin, "The Critique of Science Becomes Academic," *Science, Technology, & Human Values* 18 (1993): 247–59, but see also his *Scientific Knowledge in Controversy: The Social Dynamics of the Fluoridation Debate* (Albany: SUNY Press, 1991); "Sticking a Needle into Science: The Case of Polio Vaccines and the Origin of AIDS," *Social Studies of Science* 26 (1996): 245–76; and with Evelleen Richards, "Scientific Knowledge, Controversy, and Public Decision Making," in *Handbook*, ed. Jasanoff et al., 506–31. In the latter, the authors argue for an "integrated" approach to controversy analysis that opens up policy debates to public participation and that includes scientists, not as "privileged" experts, but rather as "partisan participants," thereby "democratiz[ing] the debate," p. 525. On "constructive technology assessment," see Arie Rip, Thomas J. Misa, and Johan Schot, eds., *Managing Technology in Society: The Approach of Constructive Technology Assessment* (London: Cassell, 1995). Eduardo Aibar, "Technological Frames in a Town Planning Controversy: Why We Do Not Have to Drop Constructivism to Avoid Political Abstinence," *Research in Philosophy and Technology* 15 (1995): 3–20. Steven Yearly, "The Environmental Challenge to Science Studies," in *Handbook*, ed. Jasanoff et al., 457–79, quotations, 477–78.

28. Edge, "Reinventing the Wheel," 12.

29. Paul Durbin, *Social Responsibility in Science, Technology, and Medicine* (Bethlehem, Pa.: Lehigh University Press, 1992), quotations, 7, 36.

30. My own involvement for several years, including two as chair, with the Pennsylvania Power and Light Public Advisory Committee was very revealing of both the possibilities and limitations of such forms of public participation in science and technology decision making. For a summary, see Eleanor Winsor and

Stephen H. Cutcliffe, "Public Involvement in Corporate Technology Decision-Making: The Case of Pennsylvania Power and Light," in *Involving the Public in Energy Facility Planning: The Electric Utility Experience*, ed. Dennis W. Duscik (Boulder, Colo.: Westview Press, 1986), chapter 13.

31. Durbin, *Social Responsibility*, 139, 172, 199.

32. Richard Sclove, *Democracy and Technology* (New York: Guilford, 1995). The Loka Institute, located in Amherst, Massachusetts, publishes a periodic electronic newsletter, the *Loka Alert*, which can be accessed on-line at: <www.loka.org>. For Barber's ideas on "strong democracy," see *Strong Democracy: Participatory Politics for a New Age* (Berkeley: University of California Press, 1984).

33. Sclove, *Democracy and Technology*, 25, 68, 206, 244.

34. Information on science shops can be found in a number of places, including Rolf Zaal and Loet Leydesdorff, "Amsterdam Science Shop and Its Influence on University Research: The Effects of Ten Years of Dealing with Non-Academic Questions," *Science and Public Policy* 14 (December 1987): 310–16; and Richard Sclove, "STS on Other Planets," *EASST Review* 15 (June 1996): 3–7, which is reprinted in Cutcliffe and Mitcham, eds., *Visions of STS*. The Loka Institute (www.loka.org) has published a lengthy report, "Community-Based Research in the United States," comparing U.S. community-based research efforts, of which there is a surprising amount, with those in the Netherlands, which are more extensive relative to population. Newsletter information about indigenous knowledge resource centers is available from the Center for International Research & Advisory Networks (CIRAN), PO Box 29777, 2502 LT The Hague, The Netherlands, or through their web site: <www.nufficcs.nl/ciran/ikdm/>. Specific details and guidelines for the Canadian CURA project can be found on their web site: <www.198.96.3/english/resnews/pressreleases/curawinners.html>.

35. On consensus conferences, see Simon Ross and John Durant, eds., *Public Participation in Science: The Role of Consensus Conferences in Europe* (London: Science Museum, 1995); Richard E. Sclove, "Democratizing Science and Technology in the Next Millennium," Plenary lecture, March 6, 1998, Annual NASTS Conference, Naperville, Ill.; Sclove, "Better Approaches to Science Policy," *Science* 279 (27 February 1998): 1283; and Sclove, "STS on Other Planets." Announcements, including e-mail and web site addresses for further information, about various nations' experiments with citizen consensus conferences can be found in *Loka Alert* 6, no. 1 (February 19, 1999) at <www.loka.org>. Brief reports on the 1999 Danish consensus conference on genetically engineered foods written by American observer Phil Bereano can be found in *Loka Alert* 6, no. 2 (May 20, 1999) and on the Loka web site, with the longer full report of the panel itself in English on the Danish Board of Technology web site:<http://www.tekno.dk/eng/publicat/genfoods.html>.

36. A report on this first U.S. citizens' advisory panel experiment, including the final "Consensus Statement," is available through the Loka Institute web site: <www.loka.org>.

37. The first POSTI workshop, entitled "Technological Innovation in a Sustainable Perspective," was held May 29–30, 1999, in Lausanne, Switzerland, in conjunction with the ESST Annual Scientific Conference. Information about the POSTI project can be found at its web site: <www.esst.uio.no/posti/>.

38. For a somewhat more extended discussion of these two examples based on the 1992 trip, see Stephen H. Cutcliffe, "Some Impressions of Science, Technology, and Society Studies in China," *Technology in Society* 15 (1993): 243–51.

39. Li Bocong, "STS in China," *Science, Technology & Society Curriculum Newsletter* 104 (Summer 1995): 1–5; Yin Deng-xiang, "STS Related Education in China," *Science, Technology & Society Curriculum Newsletter* 85 (September 1991): 11; and Wei Hongsen, "The Development of Science, Technology, and Society at Tsinghua University, Beijing, China," *Science, Technology & Society Curriculum Newsletter* 91 (September 1992): 6–8.

40. To be fair, Chinese STS scholars are not unaware of these issues as reflected both in their presentations at the "Humanistic Factors" conference and by some of the research projects in which STS faculty and graduate students engage. China also has very strict laws, even including the death penalty, for anyone building on farmland without the proper set of difficult-to-get permits. Nonetheless, there does tend to be an equation of social progress with ever more science and technology. For examples of some of the sorts of research conducted at Tsinghua University, see Cutcliffe, "Some Impressions of Science, Technology, and Society Studies in China," 246. On recent issues related to infrastructure and highway building, see Rena Singer (Knight Rider), "Sprawled Out," *The Spokesman-Review* (Spokane, Wash.) February 12, 1999, A2.

41. Quotation taken from the Lehigh University mission statement, "The Lehigh Plan," 1994.

42. On the general theme of multiple ethical issues in design that focuses on the role of the moral agent, see Caroline Whitbeck, "Ethics as Design: Doing Justice to Moral Problems," *Hastings Center Report* 26, no. 3 (1996): 9–16.

43. I do not want to leave the impression that engineers and engineering educators pay no attention to the question of societal concerns or environmental ethics, for a number certainly have, and in this regard, by way of example, I would mention the work of Joseph R. Herkert, *Social, Ethical, and Policy Implications of Engineering: Selected Readings* (New York: IEEE Press, 2000); Stephen Unger, *Controlling Technology: Ethics and the Responsible Engineer*, 2d ed. (New York: John Wiley, 1994); Mike W. Martin and Roland Schinzinger, *Ethics in Engineering*, 2d ed. (New York: McGraw-Hill, 1989); Stephen Johnston et al., *Engineering and Society: An Australian Perspective* (Pymble, Aust.: Harper Educational, 1995), especially chapter 8; John R. Wilcox and Louis Theodore, eds., *Engineering and Environmental Ethics: A Case Study Approach* (New York: John Wiley, 1998); Alistair P. Gunn and Aarne Vesiland, eds., *Environmental Ethics for Engineers* (Chelsea, Mich.: Lewis Publishers, 1986); and Vesilind and Gunn, *Engineering, Ethics, and the Environment* (New York: Cambridge University Press, 1998). See also Robert A. Frosch, "The Industrial Ecology of the 21st Century," *Scientific American* 273 (September 1995): 178; and Goldman et al., *Agile Corporations and Virtual Organizations*, especially 39, 63–64, 330, 351–52, which recognizes the importance of "green" products and processes as part of its approach to manufacturing competitiveness. I also do not mean to leave the impression that I am promoting the idea of some sort of ideal, "static" ecological balance, in contrast to the increasingly accepted notion of a "dynamic" changing ecology, but I do mean to suggest that such change ought not to be held hostage to "unthinking" technological and economic development.

44. RPI Product Design and Innovation brochure and college catalog. The UVA design effort was noted by William McDonough, Dean of Architecture, in his Joseph Needham Memorial Lecture, "Ecological Ethics and the Design Revolution," at the joint Society for the History of Technology and Society for the Social Studies of Science meeting in Charlottesville, Va., October 22, 1995. Colorado School of Mines Undergraduate Bulletin, 1998–99, 1; and "Profile of the Colorado School of Mines Graduate," 1994. I am indebted to Carl Mitcham for information with respect to Pennsylvania State University.

45. Environmental Commission, Institute of Civil Engineers of Spain, "Decalogue of the Engineer in Regard to the Natural Environment," in Carl Mitcham, *Engineering Ethics throughout the World: Introduction, Documentation, and Bibliography* (University Park, Pa.: STS Press, 1992).

46. Preamble, IEA Code of Ethics, reproduced in Johnston et al., *Engineering and Society*, 444–45.

47. For a brief history of the emergence of engineering ethics as a field of study, see Vivian Weil, "The Rise of Engineering Ethics," *Technology in Society* 6 (1985): 341–45 and "Ethics in Engineering Curricula," *Research in Philosophy and Technology* 8 (1985): 243–50, but also see Carl Mitcham, "Industrial and Engineering Ethics: Introductory Notes and Annotated Bibliography," *Research in Philosophy and Technology* 8 (1985): 251–56.

Conclusion

. . . a land ethic changes the role of Homo sapiensfrom conqueror of the land-community to plain member and citizen of it. It implies respect for the community as such. . . .
—Aldo Leopold, "The Land Ethic," 1949

The foregoing analysis has outlined the ways in which a generation of academic scholars has effected the development of a new area of interdisciplinary study devoted to the analysis of the relationships among scientific ideas, technologically based machines and activities, and societal and cultural values. It includes as well policy analysis and the activist pursuit of solutions to the perceived "problematic" nature of science and technology with their attendant societal engagements. During the course of somewhat over three decades, STS has moved from a set of disciplinary perspectives or examinations of science and technology to a more holistic, interdisciplinary understanding of the tightly woven fabric of science-technology-society. As such, it has come to recognize, indeed, it has absorbed as a central tenet of its interpretative framework, the socially constructed natures of science and technology, with their inextricably linked implications for that self-same society.

Certainly STSers have identified many interpretative hues regarding the set of intertwined relationships—ranging at the extremes from strongly deterministic to radically relativistic, and from wildly optimistic to darkly pessimistic—but for the vast majority of scholars and activists alike, STS is viewed as a place to examine science-technology-society issues. It has become, to reiterate, as David Hess has put it, "a site for public debates on issues of importance."[1] That is, STS seeks not so much to provide *the* answer, but rather to provide the space and a framework to discuss and work toward more democratic participatory solutions acceptable to all persons involved. At its most complete, STS strives to describe the ways by which scientists and engineers create and then apply their knowledge bases, all the while remaining cognizant of the implications of the resulting science and technology for society.

In this regard, STS as a whole should strive to transcend any tendency toward simplistic divisions between "High Church" accumulations of scholarly case studies, which Susan Cozzens has called "STS Thought," as an end unto itself, and "Low Church" activism, especially if uninformed by academic inquiry. Surely individuals will focus their attention in particular ways, but at its best and most inclusive, STS should seek to integrate in "ecumenical" fashion its academic scholarship with its political analysis and public activism. Robin Williams and David Edge have suggested just such a "Broad Church" approach, without insisting on a mandated "orthodoxy" as being the most beneficial, including as it would both the descriptive and the prescriptive. Here the argument is not to subsume either one within the other, but rather to maintain a creative "tension"—that is, a constructive "relationship between fact and value." So, just as the splendorous tensions of a medieval vaulted cathedral are held in place by its flying buttresses, so too should STS encompass synergistically the creative tensions of its constituent parts by an interdisciplinary array of dynamically interactive disciplinary pillars.[2]

It would be premature to argue too strongly for any sort of absolute convergence within STS that would suggest movement toward a new transdiscipline. At the same time, there do seem to be sufficient points of common agreement to suggest the outlines of a conceptual framework that entails interdisciplinary integration involving common problem solving, borrowing across disciplines, and increased methodological consistency. All of this is supported by the emergence of an institutional framework of professional and activist organizations, conferences, and participatory venues, and scholarly and educational journals all devoted to STS themes and issues.

While in many ways still limited, and assuredly in need of further refinement, there are, nonetheless, at least four interlinked concepts, and a number of widely shared methodological approaches, that transcend disciplinary vantage points alone and serve to constitute a core body of theoretical STS knowledge and practice.

1. *Constructivism.* First and foremost, STS assumes scientific and technological developments to be socially constructed phenomena. That is, science and technology, including the *content* of the former, are inherently human, and hence value-laden, activities which are always approached and hence understood through our senses. This does not deny the "constraining" order of nature, but it does entail a recognition that our *understanding* of nature and our development of technology are socially mediated processes.

2. *Contextualism.* Because science and technology are socially constructed, it follows as a corollary that they are historically, politically, and

culturally embedded, which in turn means they can only be understood *in context*. To do otherwise would be to deny their socially constructed nature. This does not deny "reality," but it does imply, as Sandra Harding has put it, that there can be so single perfect scientific map of "reality," only a series of different contextualized ways of knowing, some of which are "less false" than others.[3] Likewise, any given technological solution to a problem must be seen as contextualized within the particular socio-political-economic framework that engendered it.

3. *Problematization*. The STS view of scientific knowledge and especially technological development as value-laden, and hence nonneutral, leads to the "problematization" of both. In this view science and technology are seen as having implications, frequently positive, but often negative, for at least some portions of society. As a result, STS posits the idea that it is not only acceptable but, in fact, necessary to question the essence of scientific knowledge and the makeup and application of technological artifacts and processes with an eye toward evaluative prescription.

4. *Democratization*. Given the "problematic" natures of science and technology and accepting their "construction" by society leads to a final core idea, that of enhanced democratic control of technoscience. Here STS argues that because of the inherent implications, there need to be more explicit participatory mechanisms for enhancing public involvement in the shaping and control of science and technology, especially very early on in the decision-making processes. As Steven Goldman has put it, there should be "no innovation without representation."[4] The very fact that science and technology are seen as constructed phenomena, rather than deterministic, autonomous forces, is what opens up the possibility of this line of argument, especially for any sort of democratically organized society. Here the goal is to structure science and technology in ways that are collectively the most democratically beneficial for society. In this regard, education at both the K–12 and collegiate levels has an extremely important role to play in enhancing socially informed scientific and technological literacy in the broadest sense.

In adopting such a theoretical framework for the descriptive analysis and prescriptive evaluation of technoscience, STS can serve as a location for the discussion, both written and oral, of key societal issues of concern to democratic publics. STS also offers a "toolbox" of methodological approaches that transcend strict disciplinary bounds.[5] Thus, in addition to a basic constructivist approach, STSers frequently adopt feminist approaches such as gender analysis. Similarly, antiracist and postcolonial perspectives that reveal alternative viewpoints, especially those that involve a comparative framework, have become increasingly widespread. Observational and "ethnographic" studies, techniques initially borrowed from anthropology but that now transcend the boundaries of that field, have also proved useful. Other notions such as that of "sustainability" have done much to reconceptualize the way activist STSers think about science and technology, especially with regard to environmental issues.

Encasing this set of methodological tools is a self-reflexive recognition that STS as an approach to understanding science and technology and resolving technoscientific issues is itself societally embedded and hence equally as value-laden as its subject matter. Nonetheless, the very fact that STS in the main argues for a participatory democratic framework within which to work out decisions regarding the future course of science and technology is, in part, a reflection of just such a recognition, as well as a mechanism for helping to mitigate potential STS biases.

Taken as a whole, STS offers a framework and a variety of societally beneficial insights, if we are intelligent enough to apply them wisely toward the construction of a better world. I began this volume by quoting Emerson and Muir on the interconnectedness of life. Very early on they had recognized the intricately woven webs of ideas, machines, and values that constitute the modern world. If anything, their recognition is even more apropos at the turn of the twenty-first century. Given that recognition, I would like to conclude by drawing on the advice of yet another important environmentalist, that of Aldo Leopold.

A little more than a half-century ago, Leopold, one of America's founding scientific ecologists, wrote what has since become one of the most influential essays for the environmental movement. Entitled "The Land Ethic," and published posthumously as part of his book, *A Sand County Almanac*, the essay called for a rethinking of how we view and treat the earth. Leopold sought an extension of ethics beyond the traditional considerations focused on individuals' actions and their relations to the rest of society (important as they continue to be) to include "the land," by which he meant all of nature including animals and plants as well as the earth itself. In his words, "The land ethic simply enlarges the boundaries of the community to include soils, waters, plants, and animals, or collectively: the land." In short, he had recognized the interconnectedness of the world.

Leopold went on to argue that we need to cast off as a fallacy "the belief that economics determines all land-use." He believed we must

> . . . quit thinking about decent land use as solely an economic problem. Examine each question in terms of what is ethically and esthetically right, as well as what is economically expedient. A thing is right when it tends to preserve the integrity, stability, and beauty of the biotic community. It is wrong when it tends otherwise.

At the same time, Leopold was no neo-Luddite; in fact, he was quite the opposite. He concluded his essay as follows:

> By and large, our present problem is one of attitudes and implements. We are remodeling the Alhambra with a steamshovel, and we are proud of our yardage. We shall hardly relinquish the shovel, which after all has many good points, but we are in need of more objective criteria for its successful use.[6]

This is good advice; it is both timely and timeless. Extending Leopold's ethic to include all the socio-political-economic issues associated with technoscience, as well as with the biotic community, would hold paramount the well-being of both society and the environment.

Collectively we need to identify societally viable and environmentally sustainable responses to a wide variety of contemporary technoscientific issues. STS can provide an analytical framework and a locus of debate for doing just that. Such is at once the hope for STS and the opportunity for its greatest application.

Notes

1. David Hess, *Science Studies: An Advanced Introduction* (New York: New York University Press, 1997), 155.

2. Susan Cozzens, "The Disappearing Disciplines of STS," *Bulletin of Science, Technology & Society* 10, no. 1 (1990): 1–5; Robin Williams and David Edge, "British Perspectives on the Social Shaping of Technology: A Review of Research," in *Similar Concerns, Different Styles? Technology Studies in Western Europe*, ed. Tarja Cronberg and Knut H. Sorensen (Brussels, Luxembourg: European Commission, 1995), 274; and David Edge, "Reinventing the Wheel," in *Handbook of Science and Technology Studies*, ed. Sheila Jasanoff, et al. (Thousand Oaks, Calif.: Sage, 1995), 12.

3. See the discussion of Harding and her notion of "strong objectivity" in chapter 3 in this volume.

4. Steven L. Goldman, "No Innovation without Representation: Technological Action in a Democratic Society," in *New Worlds, New Technologies, New Issues*, ed. Stephen H. Cutcliffe, Steven L. Goldman, Manuel Medina, and José

Sanmartín (Bethlehem, Pa.: Lehigh University Press, 1992), 148–60. Langdon Winner has subsequently adopted this same phrase. See, for example, "Artifact/Ideas and Political Culture," *Whole Earth Review* 73 (Winter 1991): 18–24, also reprinted in Albert Teich, ed., *Technology and the Future*, 7th ed. (New York: St. Martin's Press 1997), 289–99.

5. Here I have borrowed the term *toolbox* from Sandra Harding who admittedly uses it in a somewhat altered context to describe the value of different local traditions and cultural conceptualizations of technoscientific knowledge. See her *Is Science Multicultural? Postcolonialisms, Feminisms, and Epistemologies* (Bloomington: Indiana University Press, 1998), 20, 190.

6. Aldo Leopold, "The Land Ethic," *A Sand County Almanac* (New York: Oxford University Press, 1949), 204, 224–26. Earlier versions of Leopold's thinking about a land ethic appeared in the 1930s as a "Conservation Ethic" and a "Biotic View of the Land"; see Susan L. Flader, *Thinking Like a Mountain: Aldo Leopold and the Evolution of an Ecological Attitude toward Deer, Wolves, and Forests* (Columbia: University of Missouri Press, 1974), chapter 1, especially 34.

Bibliographic Essay

Science, Technology, and Society Studies is oftentimes a difficult field for the new arrival to approach because of its interdisciplinarity and hence multiple perspectives. In what follows I offer a brief bibliographic overview of the field by identifying approximately 250 English-language book titles with which scholars or aspiring graduates would find it useful to familiarize themselves. The specific number is clearly minimal, for the literature is at least an order of magnitude larger, and surely some scholars will find my choices arbitrary, preferring instead their own favorites. Others will no doubt disagree with the arrangement by which I have organized them, which is further complicated by the fact that many titles might well fall into more than one category. Nonetheless, I have found this preliminary list of books, many but not all of which also appear in the notes to this volume, useful for assigning to graduate students as an introduction to the STS field. In that regard, I have chosen many of the titles because their reference notes and bibliographies lead to additional important works in specific areas of STS. I have also elected to provide a straight bibliographical listing of a core 100 titles drawn from this longer essay for those wishing a less extensive introduction.

General Introductions

The most succinct and up-to-date introduction to STS is anthropologist David J. Hess's *Science Studies: An Advanced Introduction* (New York: New York University Press, 1997), which covers some of the same ground as this volume, often in greater analytical depth with regard to the scholarly literature, but which does not deal to any great extent with the societal context that gave rise to STS nor its institutional framework. Together, Hess's book and this volume offer a useful overview of the field. A complementary survey written from more of a policy perspective is Andrew Webster's *Science, Technology, and Society* (New Brunswick, N.J.: Rutgers University Press, 1991). An older, but still useful perspective is contained in John Ziman, *An Introduction to Science Studies: The Philosophical and Social Aspects of Science and Technology* (Cambridge: Cambridge University Press, 1984). Finally, Stephen H. Cutcliffe and Carl Mitcham, eds., *Visions of STS: Contextualizing Science, Technology, and*

Society Studies (Albany: SUNY Press, 2000) offers a series of ten brief essays by different contributors who outline their perspectives on what the field of STS is and should entail.

Several volumes are revealing of the institutional organization of STS studies: Lars Fuglsang, *Technology and New Institutions: A Comparison of Strategic Choices and Technology Studies in the United States, Denmark and Sweden* (Copenhagen: Academic Press, 1993); Tarja Cronberg and Knut H. Sorenson, eds., *Similar Concerns, Different Styles? Technology Studies in Western Europe* (Luxembourg: European Commission, 1995); and Paul Wouters, Jan Annerstedt, and Loet Leydesdorff, *The European Guide to Science, Technology, and Innovation Studies* (Luxembourg: European Commission, 1999).

There are several interdisciplinary handbooks and bibliographic guides to the STS field with which one should also be familiar. From the perspective of the sociology of science and technology, including some policy perspectives, the place to start is with Sheila Jasanoff, Gerald E. Markle, James C. Petersen, and Trevor Pinch, eds., *Handbook of Science and Technology Studies* (Thousand Oaks, Calif.: Sage, 1995). The *Handbook* was intended as a replacement for Ina Spiegel-Rosing and Derek de Solla Price's *Science, Technology, and Society: A Cross-Disciplinary Perspective* (London: Sage, 1977), which is still a useful introduction to the issues and relevant literature for the early years of STS. The new *Handbook* contains over two dozen interpretative essays and an extended bibliography of some 2,400 titles drawn from the notes to the essays. As a result, the bibliography reflects the sociological and policy orientations of the volume as a whole.

Although in need of an update, one should not ignore Paul Durbin's edited *Guide to the Culture of Science, Technology, and Medicine*, 2d ed. (New York: Free Press, 1984), which includes interpretative essays and bibliographies that focus on history and philosophy, as well as sociology and policy. In addition, it covers the field of medicine, which the two previous guides largely ignore, as well as science and technology. Another collection of useful essays taking a less academic approach to STS can be found in Julian Simon, ed., *The State of Humanity* (Cambridge: Blackwell, 1995). Volume 5 of *The Reader's Advisor*, titled *The Best in Science, Technology, and Medicine* (New Providence, N.J.: R. R. Bowker, 1994), co-edited by Carl Mitcham and William F. Williams, contains a number of bibliographic chapters dealing with STS themes and issues, as well as those that treat the medical, science, and engineering fields more directly. David

McGee et al.'s introductory *Science in Society: An Annotated Guide to Resources* (Toronto: Wall & Thompson, 1989), although now out of print, is divided into four helpful sections: the nature of science, the nature of technology, humans in the environment, and current issues in science.

Several STS textbooks, especially useful in the teaching of STS issues at the undergraduate level should be mentioned: Rudi Volti, *Society and Technological Change,* 3d ed. (New York: St. Martin's Press, 1995); Martin Bridgstock et al., *Science, Technology and Society: An Introduction* (Cambridge: Cambridge University Press, 1998); Robert McGinn, *Science, Technology, and Society* (Englewood Cliffs, N.J.: Prentice Hall, 1991); and Ron Westrum, *Technologies and Society: The Shaping of People and Things* (Belmont, Calif.: Wadsworth, 1991) now out of print, but available through the author. Especially pedagogically useful is Albert H. Teich's edited selection of essays and book extracts, *Technology and the Future* (New York: St. Martin's Press, 2000), now in its eighth edition. There are also a number of helpful science-technology-medicine "encyclopedias" of various sorts, but the most avowedly STS in orientation is Rudi Volti, ed., *The Facts on File Encyclopedia of Science, Technology, and Society* (New York: Facts on File, 1999). Finally, two more popular volumes attempt to describe the natures of science and technology for an educated public: Harry Collins and Trevor Pinch, *The Golem: What You Should Know about Science*, 2d ed. (Cambridge: Cambridge University Press, 1998) and *The Golem at Large: What You Should Know about Technology* (Cambridge: Cambridge University Press, 1998).

Field-Specific Introductions: Philosophy, Sociology, History

Given its historical development and the continued influence of disciplinary perspectives on STS, I have found a number of field-specific introductions useful. For the philosophy of science, see Robert Klee, *Introduction to the Philosophy of Science: Cutting Nature at Its Seams* (Oxford: Oxford University Press, 1997); George Couvalis, *The Philosophy of Science: Science and Objectivity* (Thousand Oaks, Calif.: Sage, 1997); and Steve Fuller, *Philosophy of Science and Its Discontents*, 2d ed. (New York: Guilford Press, 1993), and for the philosophy of technology, see Carl Mitcham, *Thinking through Technology: The Path between Engineering and Philosophy* (Chicago: University of Chicago Press, 1994); Don Ihde, *Philosophy of Technology: An Introduction* (New York: Paragon House,

1993); and Joseph C. Pitt, *Thinking about Technology: Foundations of the Philosophy of Technology* (New York: Seven Bridges Press, 2000).

In the sociology of science, one should not overlook Robert K. Merton's collection of essays, *The Sociology of Science* (Chicago: University of Chicago Press, 1973), but Stephen Cole's more recent *Making Science: Between Nature and Society* (Cambridge: Harvard University Press, 1992) is more useful with regard to current STS issues. In the absence of a brief overview, the classic introduction to the sociology of technology is still Wiebe Bijker, Trevor Pinch, and Thomas P. Hughes, eds., *The Social Construction of Technological Systems* (Cambridge: MIT Press, 1987). A companion volume that focused on the issue of "closure" in technological development was Wiebe Bijker and John Law, eds., *Shaping Technology/Building Society* (Cambridge: MIT Press, 1992).

For the history of science, Helge Kragh's *The Historiography of Science* (Cambridge: Cambridge University Press, 1987) and Robert C. Olby et al.'s *Companion to the History of Modern Science* (London: Routledge, 1990) are useful introductions to historiographic traditions and interpretive trends. Jan Golinski shows how constructivist methodologies developed by sociologists and philosophers have affected the history of science in *Making Natural Knowledge: Constructivism and History of Science* (Cambridge: Cambridge University Press, 1998). Philip Kitcher's *Advancement of Science: Science without Legend, Objectivity without Illusions* (Oxford: Oxford University Press, 1993) offers a philosophical analysis of the history of science dealing with the "myths" of scientific knowledge. For the history of technology, the best introduction is still John Staudenmaier's *Technology's Storytellers: Reweaving the Human Fabric* (Cambridge: MIT Press, 1985). Stephen H. Cutcliffe and Robert C. Post, eds., *In Context: History and the History of Technology—Essays in Honor of Melvin Kranzberg* (Bethlehem, Pa.: Lehigh University Press, 1989), contains several essays that examine the field as it has become increasingly contextual in its approach.

Early Studies: Con and Pro

Among the more important early critical studies that contributed to an issue-oriented STS, were the following classics: Jacques Ellul's *The Technological Society* (French original, 1954), translated by John Wilkinson (New York: Knopf, 1964); Lewis Mumford's earlier, and somewhat more optimistic, *Technics and Civilization* (New York: Harcourt and Brace Co., 1934) and his later, more critical *The Myth of the Machine*. Vol. 1: *Technics and Human Development*; Vol. 2: *Pentagon of Power* (New York: Harcourt, Brace Jovanovich, 1967, 1970); Theodore Roszak's *The Making of a Counter Culture: Reflections on the Technocratic Society and Its Youthful Opposition* (Garden City, N.Y.: Doubleday, 1969) and *Where the Wasteland Ends: Politics and Transcendence in Postindustrial Society* (Garden City, N.Y.: Doubleday, 1972); Paul Erlich's *The Population Bomb* (New York: Ballantine Books, 1968); E. F. Schumacher's *Small Is Beautiful: Economics as if People Mattered* (New York: Harper and Row, 1973); Ivan Illich's *Tools for Conviviality* (New York: Harper and Row, 1973) and *Medical Nemesis: The Expropriation of Health* (New York: Pantheon, 1976); and Alvin Toffler's *Future Shock* (New York: Random House, 1970). Among the more valuable subsequent studies that picked up on similar themes are David Dickson's *The New Politics of Science*, 2d ed. (Chicago: University of Chicago Press, 1988); Langdon Winner's *Autonomous Technology: Technics Out-of-Control as a Theme in Political Thought* (Cambridge: MIT Press, 1977), which analyzed and extended some of the thinking of Ellul; Winner's subsequent *The Whale and the Reactor* (Chicago: University of Chicago Press, 1986); and Neil Postman's more popularized *Technopoly: The Surrender of Culture to Technology* (New York: Knopf, 1992).

For works that take a more upbeat attitude toward science and technology, one can begin with C. P. Snow's now famous "two cultures" essay, first published in 1959 as *The Two Cultures and the Scientific Revolution*, but then updated as *The Two Cultures: And a Second Look* (Cambridge: Cambridge University Press, 1964). Jacob Bronowski's *The Ascent of Man* (Boston: Little, Brown, 1976) offers an optimistic historical look at humanity's progressive attempts to control nature from agriculture to relativity. One should also consider Bronowski's *Science and Human Values*, rev. ed. (New York: Harper and Row, 1965). Emmanuel Methene's *Technological Change: Its Impact on Man and Society* (Cambridge: Harvard University Press, 1970) is based on his experience as director of the Harvard Program on Technology and Society. Mesthene is generally regarded as a conservative spokesperson in favor of technology's potential for good rather than for harm. A then contemporaneous volume written in response to some of the early criticisms of technology is civil engineer Samuel C. Florman's *The Existential Pleasures of Engineering*, rev. ed. (New York: St. Martin's Press, 1994). The new edition includes the complete text of the original work along with selected chapters from several of Florman's more recent books.

On Engineering

Other insightful works beyond those of Florman that have focused on engineering as distinct from technology include: Walter Vincenti's *What Engineers Know and How They Know It: Analytical Studies from Aeronautical History* (Baltimore: Johns Hopkins University Press, 1990); Eugene S. Ferguson's *Engineering and the Mind's Eye* (Cambridge: Cambridge University Press, 1992); David P. Billington's historical overview *The Innovators: The Engineering Pioneers who Made America Modern* (New York: Wiley, 1996); Rudi Volti's *The Engineer in History* (New York: Peter Lang, 1993); a number of works by Henry Petroski, including especially his *Invention by Design: How Engineers Get from Thought to Thing* (Cambridge: Harvard University Press, 1996); and Louis L. Bucciarelli's *Designing Engineers* (Cambridge: MIT Press, 1994). A related title of interest that focuses on the societal dimensions is Robert Pool's *Beyond Engineering: How Society Shapes Technology* (New York: Oxford University Press, 1997). A valuable collection of essays on the philosophy of engineering is contained in Paul T. Durbin, ed., *Critical Perspectives on Nonacademic Science and Engineering* (Bethlehem, Pa.: Lehigh University Press, 1991).

Philosophy of Science and Technology

Philosophers of science and technology have raised important contextual questions regarding the structure of scientific knowledge and the development of technology. A common and helpful starting point is Thomas Kuhn's *The Structure of Scientific Revolutions*, 2d ed. (Chicago: University of Chicago Press, 1970), which moved away from more traditional positivist views of science to incorporate a more contextual, even if not relativist, interpretation. Other useful philosophical discussions of epistemological issues include the work of Larry Laudan, *Progress and Its Problems: Toward a Theory of Scientific Growth* (Berkeley: University of California Press, 1977); *Science and Values* (Berkeley: University of California Press, 1984); and *Science and Relativism: Some Key Controversies in the Philoso-*

phy of Science (Chicago: University of Chicago Press, 1990); Ronald N. Giere, *Explaining Science: A Cognitive Approach* (Chicago: University of Chicago Press, 1990); Steve Fuller, *Social Epistemology* (Bloomington: Indiana University Press, 1988); *Philosophy, Rhetoric, and the End of Knowledge: The Coming of Science and Technology Studies* (Madison: University of Wisconsin Press, 1993); and *Science* (Minneapolis: University of Minnesota Press, 1997); and Joseph Rouse, *Knowledge and Power: Toward a Political Philosophy of Science* (Ithaca, N.Y.: Cornell University Press, 1987) and *Engaging Science: How to Understand Its Practices Philosophically* (Ithaca, N.Y.: Cornell University Press, 1996).

Important works in the philosophy of technology include Albert Borgmann's *Technology and the Character of Contemporary Life* (Chicago: University of Chicago Press, 1987), where he distinguishes between the domination of technological "devices" and "focal things" that tend more to combine ends and means, and his more recent *Crossing the Postmodern Divide* (Chicago: University of Chicago Press, 1992); phenomenologist Don Ihde's sequence of three volumes, *Technics and Praxis* (New York: Kluwer, 1979); *Existential Technics* (Albany: SUNY Press, 1983); *Technology and the Lifeworld: From Garden to Earth* (Bloomington: Indiana University Press, 1990); Hans Jonas's *The Imperative of Responsibility: In Search of an Ethics for the Technological Age* (Chicago: University of Chicago Press, 1985) in which he calls for "foresight and responsibility"; and Andrew Feenberg's post-Marxist *Critical Theory of Technology* (New York: Oxford University Press, 1991), which argues in support of worker-controlled society. Feenberg and Alistair Hannay's edited volume *Technology and the Politics of Knowledge* (Bloomington: Indiana University Press, 1995) also contains a number of valuable essays covering a sampling of important current themes in the philosophy of technology, touching among other things on the Frankfurt school, Heidegger, Marcuse and Habermas, social construction of technology, media theory, and feminist perspectives. Carl Mitcham and Robert Mackey's edited collection, *Philosophy and Technology: Readings in the Philosophical Problems of Technology*, rev. ed. (New York: Free Press, 1983), contains a good selection of essays along with a helpful bibliography.

One area to which philosophers and other STS scholars have devoted quite a bit of attention has been that of the ethical issues surrounding science, technology, and medicine. A sampling of relevant titles dealing with general issues would include the following: Aant Elzinga et al., *In Science We Trust? Moral and Political Issues of Science in Society* (Lund,

Sweden: Lund University Press, 1990); Loren R. Graham, *Between Science and Values* (New York: Columbia University Press, 1983); William W. Lowrance, *Modern Science and Human Values* (Oxford: Oxford University Press, 1985); Ian Barbour, *Ethics in an Age of Technology* (New York: HarperCollins, 1993) and his related *Religion in an Age of Science* (New York: HarperCollins, 1990); and Kristin S. Shrader-Frechette, ed., *Technology and Values* (Lanham, Md.: Rowman & Littlefield, 1997). For ethical treatments of more specific themes, see Shrader-Frechette, *Nuclear Power and Public Policy: The Social and Ethical Problems of Fission Technology* (Boston: Kluwer Academic, 1980); Eugene C. Hargrove, *Foundations of Environmental Ethics* (New York: Prentice Hall, 1989); Louis P. Pojman, ed., *Environmental Ethics: Readings in Theory and Application* (Boston: Jones & Bartlett, 1994); Tom L. Beauchamp and James F. Childress, *Principles of Biomedical Ethics*, 4th ed. (Oxford: Oxford University Press, 1994); Leon R. Kass, *Toward a More Natural Science: Biology and Human Affairs* (New York: Free Press, 1985); Deborah Johnson, *Computer Ethics* (New York: Prentice Hall, 1985); Stephen H. Unger, *Controlling Technology: Ethics and the Responsible Engineer*, 2d ed. (New York: Wiley, 1993); and Carl Mitcham and Philip Siekevitz, eds., *Ethical Issues Associated with Scientific and Technological Research for the Military* (New York: New York Academy of Sciences, 1989).

Both the philosophy of science and the philosophy of technology have also contributed numerous studies of particular scientific issues and technological developments. An excellent introductory listing of such works can be found in Paul Durbin's "Philosophy of Science, Technology, and Medicine," chapter 4 in *The Best in Science, Technology, and Medicine*, ed. Mitcham and Williams, cited above.

History of Science and Technology

Historians of science and technology have been less inclined to write about general theoretical issues, preferring instead to focus on specific events and issues. Nonetheless, a number of works have proved useful from an STS perspective and should not be overlooked. David Hull's *Science as a Process: An Evolutionary Account of the Social and Conceptual Development of Science* (Chicago: University of Chicago Press, 1988) offers a philosophically oriented assessment based on the history of the biological research community. Robert Proctor examines challenges to claims of scientific objectivity in *Value Free Science? Purity and Power in Modern Knowledge* (Cambridge: Harvard University Press, 1991). Thomas Kuhn's *The Essential Tension: Selected Studies in Scientific Tradition and Change* (Chicago: University of Chicago Press, 1977) contains a number of his essays written during the 1960s and early 1970s that deal with historical and historiographic themes. Londa Schiebinger's *The Mind Has No Sex? Women in the Origins of Modern Science* (Cambridge: Harvard University Press, 1991) emphasizes the role of women in science. Jerome Ravetz's *Scientific Knowledge and Its Social Problems* (Oxford: Oxford University Press, 1971) is an early work exploring science as socially conditioned, while Daniel J. Kevles's *The Physicists: The History of a Scientific Community in Modern America,* 2d ed. (Cambridge: Harvard University Press, 1987) emphasizes the political dimensions of the development of the American physics community since the late nineteenth century. Historian Derek de Solla Price's quantitative sociological study, *Little Science, Big Science* (New York: Columbia University Press 1963), identified what he viewed as a potentially significant, and unsustainable, exponential increase in scientific activity, especially as reflected in numbers of publications. A particularly important case study revealing the societally grounded nature of scientific conceptualization is sociologist Steve Shapin and historian Simon Schaffer's *Leviathan and the Air-Pump: Hobbes, Boyle and the Experimental Life* (Princeton: Princeton University Press, 1985). Shapin's *A Social History of Truth: Civility and Science in Seventeenth-Century England* (Chicago: University of Chicago Press, 1994) seeks to understand scientific "truth" claims, while his *The Scientific Revolution* (Chicago: University of Chicago Press, 1996) examines the historically situated emergence of modern scientific knowledge.

A recent history of science and technology that also includes some discussion of societal implications is Keith J. Laidler's *To Light Such a Candle: Chapters in the History of Science and Technology* (Oxford: Oxford University Press, 1998). George Basalla's *The Evolution of Technology* (Cambridge: Cambridge University Press, 1988) discusses the evolutionary development of technological forms over time, while another useful overview exploring the impact of technology on society and the environment is R. Angus Buchanan's *The Power of the Machine: The Impact of Technology from 1700 to the Present Day* (New York: Viking Penguin, 1993). David F. Noble's *America by Design: Science, Technology, and the Rise of Corporate Capitalism* (New York: Alfred A. Knopf, 1977) links the professionalization of science and technology to corporate economic power and development. Michael Adas analyzes the relationship between science, technology, and colonialism in *Machines as Measures of Man: Science, Technology, and Ideologies of Western Dominance* (Ithaca, N.Y.: Cornell University Press, 1990). An early classic study showing the societal implications of such technological developments as the stirrup and crop rotation is Lynn White jr.'s *Medieval Technology and Social Change* (Oxford: Oxford University Press, 1962). Several works by Arnold Pacey are also valuable for their general depiction of the contextual nature of technological development: *The Culture of Technology* (Cambridge: MIT Press, 1983); *Technology in World Civilization: A Thousand-Year History* (Cambridge: MIT Press, 1990); *The Maze of Ingenuity: Ideas and Idealism in the Development of Technology*, 2d ed. (Cambridge: MIT Press, 1992); and *Meaning in Technology* (Cambridge: MIT Press, 1999). Two more focused studies that are particularly revealing from an STS perspective are Ruth Schwartz Cowan's *More Work for Mother: The Ironies of Household Technology from the Open Hearth to the Microwave* (New York: Basic Books, 1983) and Thomas P. Hughes's *Networks of Power: Electrification in Western Society, 1880–1930* (Baltimore: Johns Hopkins University Press, 1983) where he works out some of his ideas regarding systems thinking and the notion of technological "momentum." Also valuable is Hughes's broader treatment of U.S. technological development, *American Genesis: A Century of Invention and Technological Enthusiasm, 1870–1970* (New York: Viking, 1989), as is David E. Nye's depiction of the affection of the United States for technology, *American Technological Sublime* (Cambridge: MIT Press, 1994).

For those interested in further titles related to the history of science and technology, Henry Lowood's essay in Mitcham and Williams's bibliographic compilation noted above is a good place to begin. Also helpful is Robert K. DeKosky's "Science, Technology and Medicine," in the American Historical Association's *Guide to Historical Literature*, ed. Mary Beth Norton, 3d ed. (New York: Oxford University Press, 1995), vol. 1, sec. 4, 77–121.

Sociology of Scientific Knowledge

The sociology of scientific knowledge (SSK) and later the social construction of technology (SCOT), often referred to more generally as "social construction," has become a central feature of STS scholarship. Beginning in the mid-1970s and continuing through the 1980s a number of key studies characterized the central tenets of the SSK approach, including the role of interests and the notions of symmetry and reflexivity. Among the most important were David Bloor's *Knowledge and Social Imagery*, 2d ed. (Chicago: University of Chicago Press, 1991); Barry Barnes's *Interests and the Growth of Knowledge* (London: Routledge, 1977) and *About Science* (Oxford: Blackwell, 1985); Michael J. Mulkay's *Science and the Sociology of Knowledge* (London: George Allen & Unwin, 1979); and Harry M. Collins, *Changing Order: Replication and Induction in Scientific Practice* (Beverly Hills, Calif.: Sage, 1985). Other more recent volumes that summarize SSK developments and approaches include: Mulkay's *Sociology of Science: A Sociological Pilgrimage* (Bloomington: Indiana University Press, 1991); Steve Woolgar's *Science: The Very Idea* (London: Tavistock, 1988); Michael Lynch, *Scientific Practice and Ordinary Action: Ethnomethodology and Social Studies of Science* (Cambridge: Cambridge University Press, 1993); and Barry Barnes, David Bloor, and John Henry's *Scientific Knowledge: A Sociological Analysis* (Chicago: University of Chicago Press, 1996).

Much of the early SSK literature was published in essay form, and several edited collections appeared that brought together the best of these articles. Among the anthologies with which one should be familiar are the following: Barry Barnes and David Edge, eds., *Science in Context: Readings in the Sociology of Science* (Milton Keynes, U.K.: Open University Press; Cambridge: MIT Press, 1982); Karin Knorr Cetina and Michael J. Mulkay, eds., *Science Observed: Perspectives on the Social Study of Science* (Beverly Hills, Calif.: Sage, 1983); Steven Woolgar, ed., *Knowledge and Reflexivity: New Frontiers in the Sociology of Knowledge* (Beverly Hills, Calif.: Sage, 1988); and Michael Lynch and Steven Woolgar, eds., *Representation in Scientific Practice* (Cambridge: MIT Press, 1990). The most recent of such anthologies is Mario Biagioli, ed., *The Science Studies Reader* (New York: Routledge, 1999), which focuses on the practice of science and includes material published mostly during the 1990s.

SSK has drawn on a number of illustrative case studies including David Edge and Michael Mulkay's *Astronomy Transformed: The Emergence of Radio Astronomy in Britain* (New York: Wiley, 1976); Andrew Pickering's *Constructing Quarks: A Sociological History of Particle Physics* (Chicago: University of Chicago Press, 1984); G. Nigel Gilbert and Michael Mulkay's *Opening Pandora's Box: A Sociological Analysis of Sci-*

entists' Discourse (Cambridge: Cambridge University Press, 1984); Trevor Pinch's *Confronting Nature: The Sociology of Solar Neutrino Detection* (Dordrecht: Reidel, 1986); and Peter Galison's *How Experiments End* (Chicago: University of Chicago Press, 1987).

The Turn toward Technology

The complementary "turn toward technology" as a way of understanding technological artifacts and technological knowledge as societally shaped phenomena was introduced in Donald MacKenzie and Judith Wajcman's edited collection, *The Social Shaping of Technology: How the Refrigerator Got Its Hum* (Milton Keynes, U.K.: Open University Press, 1985), which has recently been much revised in a second edition (1999). In addition to the two edited collections by Bijker et al., noted above, his *Of Bicycles, Bakelites, and Bulbs: Toward a Theory of Sociological Change* (Cambridge: MIT Press, 1995) is particularly valuable both for the cases studies and for its theoretical framework. Other valuable case studies in this vein include: MacKenzie's *Inventing Accuracy: A Historical Sociology of Nuclear Missile Guidance* (Cambridge: MIT Press, 1990); Harry M. Collins's *Artificial Experts: Social Knowledge and Intelligent Machines* (Cambridge: MIT Press, 1992); and Stuart S. Blume's *Insight and Industry: On the Dynamics of Technological Change in Medicine* (Cambridge: MIT Press, 1992). Also insightful is a collection of MacKenzie's articles, *Knowing Machines: Essays on Technological Change* (Cambridge: MIT Press, 1996), in which he examines the way "Knowledge"—both explicit systematized knowledge and informal, tacit "know how"—is embedded in and helps to shape "high" technologies.

Ethnographic Studies

During the 1980s, the sociologists and anthropologists of science became increasingly interested in what scientists actually did within their laboratories, not just in analyzing historically what they published and said they had done. Thus emerged several "laboratory studies" that adopted anthropological observational techniques. Among the more useful have been Bruno Latour and Steve Woolgar's *Laboratory Life: The Social Construction of Scientific Facts*, 2d ed. (Princeton: Princeton University Press, 1986): Karin Knorr-Cetina's *The Manufacture of Knowledge: An Essay on the Constructivist and Contextual Nature of Science* (New York and Oxford: Pergamon, 1981); Sharon Traweek's *Beamtimes and Lifetimes: The World of High Energy Physicists* (Cambridge: Harvard University Press, 1988); and the work of Joan Fujimura, summarized most recently in her book, *Crafting Science: A Sociohistory of the Quest for Genetics of Cancer* (Cambridge: Harvard University Press, 1996). Latour subsequently generalized outward from his own research to write a broader assessment of how "technoscience" needs to be conceptualized within the notion of "actor-network theory" in *Science in Action: How to Follow Scientists and Engineers through Society* (Cambridge: Harvard University Press, 1987). Most recently, Karen Knorr Cetina has compared ethnographically the fields of high energy physics and molecular biology in *Epistemic Cultures: How the Sciences Make Knowledge* (Cambridge: Harvard University Press, 1998).

Several recent works have tried to grapple with the issue of realism and the objectivity of science without giving up the role of societal influences. Andrew Pickering, in *The Mangle of Practice: Time, Agency, and Science* (Chicago: University of Chicago Press, 1995), seeks to show how the practice of science entails a "dialectic" of "resistances" that includes the constraints of material agency and "accommodations," that is, "an active human strategy of response. . . . " Pickering's earlier edited volume, *Science as Practice and Culture* (Chicago: University of Chicago Press, 1992), contains a number of valuable essays that focus on the actual practice of doing science and the "culture" within which scientists go about doing their work. In *Uncertain Knowledge: An Image of Science for a Changing World* (Cambridge: Cambridge University Press, 1996), R. G. A. Dolby argues for a "modest realism," one that views science as a reasonably realistic representation of nature, but that recognizes that the frameworks within which scientific knowledge is formed and understood will change, hence the "uncertainty" of such knowledge. Thomas Gieryn, in his book *Cultural Boundaries of Science: Credibility on the Line* (Chicago: University of Chicago Press, 1998), attempts to map the boundaries between science and nonscience by analyzing the sociocognitive processes that distinguish them. Another recent work that uses evolutionary theory as a case study of the nature of science is philosopher of biology Michael Ruse's *Mystery of Mysteries: Is Evolution a Social Construction?* (Cambridge: Harvard University Press, 1999). In *The Advancement of Science* noted above, Philip Kitcher defends a broadly realist position as he attempts "to steer a middle course between legend and relativism" (p. 7n.7). Finally, philosopher Ian Hacking, in *The Social Construction of What?* (Cambridge: Harvard University Press, 1999), explores the meaning and significance of the idea of social construction. Collectively these studies position STS between the extremes of a positivist realism and a strong relativism, and between SSK and SCOT, which tend to privilege human agency, and that of actor network theory, which blurs distinctions between the human and the nonhuman.

Feminist and Multicultural Scholarship

During the 1990s STS scholarship also began to branch out from its earlier concentration on SSK and constructivist case studies to include societally broader critical questions that grapple with issues of gender and the postcolonial and antiracist implications of contemporary science and technology. Two important works that attempt to focus on the role of culture and the role of power in technoscientific development for our contemporary global society are anthropologist David J. Hess's *Science and Technology in a Multicultural World: The Politics of Facts and Artifacts* (New York: Columbia University Press, 1995) and philosopher Sandra Harding's *Is Science Multicultural? Postcolonialisms, Feminisms, and Epistemologies* (Bloomington: Indiana University Press, 1998). The question of race has begun to appear as a topic of inquiry in recent STS studies, and it is the focus of an edited volume by Harding entitled *The Racial Economy of Science* (New York: Routledge, 1993). Donna Haraway's *Primate Visions: Gender, Race, and Nature in the World of Modern Science* (New York: Routledge, 1989) shows how race as well as gender can affect the way scientists, in this case primatologists, adopt different methodologies, while at the same time alerting us to the possibilities of racial hierarchies even within previously excluded groups. See also her *Simians, Cyborgs, and Women: The Reinvention of Nature* (London: Routledge, 1991). A classic study dealing with environmental racism and injustice with respect to the siting of hazardous waste dumps is Robert Bullard, *Dumping on Dixie: Race, Class, and Environmental Equity* (Boulder, Colo.: Westview Press, 1990). Also useful is his *Confronting Environmental Racism: Voices from the Grass Roots* (Boston: South End, 1993).

One aspect of the cultural analysis of science and technology that has been particularly well developed during the 1980s and 1990s has been that of feminist and gender studies. Among the more important works that should be considered are those of Evelyn Fox Keller, especially her *Reflections on Gender and Science* (New Haven: Yale University Press, 1985), in which she argues both science and gender are social constructs, and a collection of her essays on the role of language in science, *Secrets of Life, Secrets of Death: Essays on Language, Gender, and Science* (New York: Routledge, 1992). Sandra Harding's *The Science Question in Feminism* (Ithaca, N.Y.: Cornell University Press, 1986) assesses various trends in feminist critiques of science, while her *Whose Science? Whose Knowledge? Thinking from Women's Lives* (Ithaca, N.Y.: Cornell University Press, 1991) argues for increased awareness of the ways biases, including gender and the role of power, affect ways of seeking and formulating scientific knowledge. Another instructive gender analysis is that of philosopher Helen E. Longino, who, in her book *Science as Social Knowledge: Values and Objectivity* (Princeton: Princeton University Press, 1990), draws on feminist theory in the area of human biological evolution to argue for a "contextual empiricism" that reconciles scientific objectivity and the shaping influence of social values. Also valuable is Ruth Hubbard's *The Politics of Women's Biology* (New Brunswick, N.J.: Rutgers University Press, 1990). Londa Schiebinger examines the historical role of women in science, including the ways feminist perspectives, as well as gender itself, have shaped the creation of scientific knowledge in *Has Feminism Changed Science?* (Cambridge: Harvard University Press, 1999).

Somewhat less in the way of generalized feminist scholarship on technology has been written than on science; however, there are a number of important works that should be considered. Perhaps the best starting point is with Judy Wajcman's *Feminism Confronts Technology* (University Park: Penn State University Press, 1991), which compares feminist theories of science and technology and summarizes much of the literature in several specific areas of research including household technology, the paid workplace, the built environment, and medical and reproductive technologies. Also useful are earlier collections of essays edited by Joan Rothschild, *Machina ex Dea: Feminist Perspectives on Technology* (New York: Pergamon, 1983) and by Chris Kramarae, *Technology and Women's Voices* (New York: Routledge, 1988). Two works by sociologist Sally Hacker are revealing of issues faced by women in the workplace: *Pleasure, Power, and Technology: Some Tales of Gender, Engineering, and the Cooperative Workplace* (Boston: Unwin Hyman, 1989) and her edited collection with Dorothy

Smith and Susan Turner, *"Doing It the Hard Way": Investigations of Gender and Technology* (Boston: Unwin Hyman, 1990).

There are a number of valuable historical studies that deal with technological themes. In addition to the work of Ruth Cowan on household technology noted previously, see also Susan Strasser, *Never Done: A History of American Housework* (New York: Pantheon, 1982). Michele Martin in *Hello Central? Gender, Technology, and Culture in the Formation of Telephone Systems* (Montreal: McGill-Queen's University Press, 1991) and L. Rakow in *Gender on the Line: The Telephone and Community Life* (Champaign: University of Illinois Press, 1992) focus on women's different relationship to the telephone than that of men, while Virginia Scharff treats the place of women within the emergence of the early car culture in America in *Taking the Wheel: Women and the Coming of the Motor Age* (Albuquerque: University of New Mexico Press, 1992). On reproductive technologies, see Robbie Davis-Floyd, *Birth as an American Rite of Passage* (Berkeley: University of California Press, 1992) and Marilyn Strathern, *Reproducing the Future: Anthropology, Kinship and the New Reproductive Technologies* (New York: Routledge, 1992).

Science and Technology Policy

Science and technology policy issues are an arena where academic intellectual interests frequently intersect with those of the more activist-oriented part of STS. Thus, STS involves the politics of managing science and technology as well as the science and technology policies actually developed within the governmental arena. For general overviews of policy issues, one should consult the following: Alexander J. Morin, *Science Policy and Politics* (Englewood Cliffs, N.J.: Prentice Hall, 1993); Patrick Hamlett, *Understanding Technological Politics: A Decision-Making Approach* (Englewood Cliffs, N.J.: Prentice Hall, 1992); and John Street, *Politics and Technology* (New York: Guilford, 1992). Bruce L. R. Smith's *American Science Policy Since World War II* (Washington, D.C.: Brookings Institute, 1990) in conjunction with David Dickson's *The New Politics of Science,* noted above, provide useful historical coverage. Smith's *The Advisors: Scientists in the Policy Process* (Washington, D.C.: Brookings Institute, 1992) and Sheila Jasanoff's *The Fifth Branch: Science Advisors as Policy Makers* (Cambridge: Harvard University Press, 1990) focus on the participation in policy making of scientists themselves.

Although somewhat older, still valuable is Daniel S. Greenberg's *The Politics of Pure Science* (New York: New American Library, 1967).

The following works offer more interpretative insights into policy issues: Anne L. Hiskes and Richard P. Hiskes, *Science, Technology, and Policy Issues* (Boulder, Colo.: Westview Press, 1986) focuses on normative concerns related to policy making; Deborah Shapley and Rustum Roy, *Lost at the Frontier: U.S. Science and Technology Policy Adrift* (Philadelphia: ISI Press, 1985) argue for greater focus on engineering and applied science, believing basic research is over funded; while the work of Edward Wenk, *Tradeoffs: Imperatives of Choice in a High-Tech World*, reprint ed. (Baltimore: Johns Hopkins University Press, 1989) and *Making Waves: Engineering, Politics and the Social Management of Technology* (Urbana: University of Illinois Press, 1995), argues for long-range management of technology in the face of inadequate policy.

For issues related to the law and policy, consult Sheila Jasanoff's *Science at the Bar: Law, Science, and Technology in America* (Cambridge: Harvard University Press, 1995). Steven Yearly's *Science, Technology, and Social Change* (Boston: Unwin Hyman, 1988) examines how science and technology relate to change in both developed and undeveloped nations. Susan E. Cozzens and Thomas F. Gieryn's edited volume *Theories of Science in Society* (Bloomington: Indiana University Press, 1990) includes a selection of essays that relate theories of science and technology to the policy-making setting. One should also consult the American Association for the Advancement of Science's annual *AAAS Science and Technology Yearbook* (Washington, D.C.: AAAS, 1991–).

The policy related literature is quite large, but a few useful, more pointed studies and collections include: Bruce Lewenstein, ed. *When Science Meets the Public* (Washington, D.C.: AAAS, 1992) and Alan Irwin and Brian Wynne, eds., *Misunderstanding Science? The Public Reconstruction of Science and Technology* (Cambridge: Cambridge University Press, 1996) on public understanding and participation in science and technology decision making; Dorothy Nelkin, *Selling Science: How the Press Covers Science and Technology*, rev. ed. (New York: W. H. Freeman, 1995) on the relationships between scientists and journalists; William Lowrance, *Of Acceptable Risk: Science and the Determination of Safety* (Los Altos, Calif.: William Kaufmann, 1976); Charles Perrow, *Normal Accidents: Living with High-Risk Technologies* (New York: Basic Books, 1984); Joseph Morone and Edward J. Woodhouse, *Averting Catastrophe: Strategies for Regulating Risky Technologies* (Berkeley: University of California Press, 1986); Deborah G. Mayo and Rachelle D. Hollander, *Acceptable Evidence: Science and Values in Risk Management* (New York: Oxford University Press, 1991); and the more methods-oriented Daniel M. Kammen and David M. Hassenzahl, *Should We Risk It? Exploring Environmental, Health, and Technological Problem Solving* (Princeton: Princeton University Press, 1999), all on technological risk and its management; Daryl Chubin and Edward J. Hackett, *Peerless Science: Peer Review and U.S. Science Policy* (Albany: SUNY Press, 1990) and Robert Bell, *Impure Science: Fraud, Compromise, and Political Influence in Scientific Research* (New York: Wiley, 1992) on peer review and misconduct as related to science policy; and Dorothy Nelkin, ed., *Controversy: Politics of Technical Decisions*, 3d ed. (Newbury Park, Calif.: Sage, 1992).

Among recent works that argue for more thoughtful democratic participation in the technoscientific decision-making process are: Charles Pillar's *The Fail-Safe Society: Community Defiance and the End of American Technological Optimism* (New York: Basic Books, 1991); Paul T. Durbin's *Social Responsibility in Science, Technology, and Medicine* (Bethlehem, Pa.: Lehigh University Press, 1992); Richard E. Sclove's *Democracy and Technology* (New York: Guilford, 1995); and Daniel Sarewitz's *Frontiers of Illusion: Science, Technology and the Politics of Progress* (Philadelphia: Temple University Press, 1996).

Environmental Issues

Another large area with a distinct literature of its own is that of environmental studies. A few books of particular interest to STS scholars beyond those already mentioned in other contexts would include the following titles. The best survey of U.S. environmental history is John Opie's *Nature's Nation: An Environmental History of the United States* (Fort Worth, Tex.: Harcourt Brace, 1998), but also still valuable for exploring changing attitudes toward nature and wilderness is Roderick Nash's, *Wilderness and the American Mind,* 3d ed. (New Haven: Yale University Press, 1982). The history of the recent U.S. environmental movement is covered in Samuel P. Hays's *Beauty, Health and Permanence: Environmental Politics in the United States, 1955–1985* (Cambridge: Cambridge University Press, 1987), while John McCormick's *Reclaiming Paradise: The Global Environmental Movement* (London: Belhaven, 1989) encompasses a broader comparative view. Other more focused historical studies worthy of note are: Carolyn Merchant's *The Death of Nature: Women, Ecology and the Scientific Revolution* (San Francisco: Harper and Row, 1980) on science and its relationship to nature; Daniel Yergin's analysis of the petroleum industry, *The Prize: The Epic Quest for Oil, Money and Power* (New York: Simon and Schuster, 1992); James Williams's *Energy and the Making of Modern California* (Akron, Ohio: University of Akron Press, 1997); and two books by Donald E. Worster, *Nature's Economy: A History of Ecological Ideas*, 2d ed. (Cambridge: Cambridge University Press, 1985) and *Rivers of Empire: Water, Aridity and the Growth of the American West* (New York: Pantheon, 1985).

Among the early STS-related studies raising questions about the implications of modern technoscientific society for the environment are: Rachel Carson's *Silent Spring* (Boston: Houghton Mifflin, 1962); Barry Commoner's *The Closing Circle* (New York: Alfred A. Knopf, 1971); Donella Meadows et al.'s computer model-based analyses of resource consumption related to population, *The Limits to Growth* (New York: Signet Books, 1972) and the comparative update, *Beyond the Limits: Confronting Global Collapse, Envisioning a Sustainable Future* (Post Mills, Vt.: Chelsea Green, 1992); and Amory Lovin's *Soft Energy Paths: Toward a Durable Peace* (New York: Harper and Row, 1977), which argued for conservation and small-scale technology, over large, centralized energy systems such as nuclear power. A more recent work in this same tradition is Bill McKibben's *The End of Nature* (New York: Random House, 1989). One should also consider the range of materials published by Lester Brown and the Worldwatch Institute located in Washington, D.C., including their annual *State of the World* and *Vital Signs* series, which monitor and assess the world's progress toward a sustainable society. Two important studies that question whether there is really the need for as much concern regarding the environment are Julian Simon's *The Ultimate Resource* (Princeton: Princeton University Press, 1981), which argues that society has in the past always found resource substitutes when shortages became severe, hence making replacements cost effective, and that there is no reason to expect the future to be any different; and Aaron Wildavsky's *But Is It True? A Citizen's Guide to Environmental Health and Safety Issues* (Cambridge: Harvard University Press, 1995).

Among economic and policy-oriented books, one should consider Herman Daly and John Cobb Jr.'s *For the Common Good: Redirecting the Economy toward Community, the Environment, and a Sustainable Future* (Boston: Beacon Press, 1989), which suggests alternative ways to value nature; David Vogel's comparative study, *National Styles of Regulation: Environmental Policy in Great Britain and the United States* (Ithaca, N.Y.: Cornell University Press, 1986); the work of Sheila Jasanoff noted previously, as well as her *Risk Management and Political Culture* (New York: Russell Sage, 1986); Sheldon Krimsky and Alonzo Plough's *Environmental Hazards: Communicating Risks as a Social Process* (Dover, Mass.: Auburn House, 1988); and Daniel J. Fiorino's *Making Environmental Policy* (Berkeley: University of California Press, 1995). See also John Byrne and Daniel Rich's edited volume, *The Politics of Energy Research and Development* (New Brunswick, N.J.: Transaction, 1986); Joseph G. Morone and Edward J. Woodhouse's *The Demise of American Nuclear Power: Learning from the Failure of a Politically Unsafe Technology* (New Haven: Yale University Press, 1989); Lynton Caldwell's *Between Two Worlds: Science, the Environmental Movement and Policy Choice* (Cambridge: Cambridge University Press, 1990); Robert C. Paelke's *Environmentalism and the Future of Progressive Politics* (New Haven: Yale University Press, 1989); Steven Yearly's study of the green movement, *The Green Case* (London: Routledge, 1992); and Franz Foltz's analysis of the politics of global climate change, *The Increasing Participation in Science: The Rise and Fall of the U.S. Global Change Research Program* (Bethlehem, Pa.: Lehigh University Press, forthcoming).

Finally, several more philosophically inclined works that raise social and ethical questions regarding nature and the environment would include: Bill Devall and George Sessions's *Deep Ecology: Living as if Nature Mattered* (Salt Lake City, Utah: Peregrine Smith Books, 1985); Mark Sagoff's *The Economy of Earth: Philosophy, Law, and the Environment* (Cambridge: Cambridge University Press, 1988); J. Baird Callicott's *In Defense of the Land Ethic* (Albany: SUNY Press, 1989); and Max Oelschlaeger's *The Idea of Wilderness from Prehistory to the Present* (New Haven: Yale University Press, 1991). In addition to the Pojman collection noted above, three other useful anthologies that include a range of philosophical and critical viewpoints are Michael Zimmerman et al.'s *Environmental Philosophy: From Animal Rights to Radical Ecology* (Englewood Cliffs, N.J.: Prentice Hall, 1993); Carolyn Merchant's *Ecology: Key Concepts in Critical Theory* (Atlantic Highlands, N.J.: Humanities Press, 1994); and Don E. Marietta and Lester Embree's *Environmental Philosophy and Environmental Activism* (Lanham, Md.: Rowman & Littlefield, 1995). Lastly, two works that argue for intensified environmental consideration with regard to science and technology are David Strong's *Crazy Mountains: Learning from Wilderness to Weigh Technology* (Albany: SUNY Press, 1995) and Alan Drengson's *The Practice of Technology: Exploring Technology, Ecophilosophy, and Spiritual Disciplines for Vital Links* (Albany: SUNY Press, 1995).

Science Wars

Although in many ways the topic of the so-called science wars is not central to the main discourse of STS, the fact that it absorbed so much attention during the second half of the 1990s requires brief mention of at least several key works in the controversy. A good place to begin is with Gerald Holton's *Science and Anti-Science* (Cambridge: Harvard University Press, 1993). For works particularly critical of STS and constructivist approaches to understanding science and technology, see Paul R. Gross and Norman Levitt, *Higher Superstition: The Academic Left and Its Quarrels with Science*, 2d ed. (Baltimore: Johns Hopkins University Press, 1998); Gross, Levitt, and Martin W. Lewis, eds., *The Flight from Science and Reason* (New York: New York Academy of Sciences, 1996); Alan Sokal and Jean Bricmont, *Fashionable Nonsense: Postmodern Intellectuals' Abuse of Science* (New York: Picador, 1998); and Noretta Koertge, ed., *A House Built on Sand: Exposing Postmodernist Myths about Science* (New York: Oxford University Press, 1998). Two collections of essays in response to the attacks of the anti-STS group include Andrew Ross, ed., *Science Wars* (Durham, N.C.: Duke University Press, 1996) and Timothy Lenoir, ed., *Inscribing Science: Scientific Texts and the Materiality of Communication* (Stanford: Stanford University Press, 1998). Ullica Segerstrale's edited collection, *Beyond the Science Wars: The Missing Discourse about Science and Society* (New Brunswick, N.J.: Rutgers University Press, 2000), includes essays that attempt to move the discussion forward. Also interesting in this regard is Bruno Latour's *Pandora's Hope: Essays on the Reality of Science Studies* (Cambridge: Harvard University Press, 1999) in which he does not deny the existence of an outside world, but at the same time argues that it makes no sense to talk of it in ahistorical, inhuman terms.

Science and Technology Education

The literature on science and technology education is vast, but among the titles relevant to STS education that should be consulted are: Dennis Cheek, *Thinking Constructively about Science, Technology and Society Education* (Albany: SUNY Press, 1992); Roger T. Cross and Ronald F. Price, *Teaching Science for Social Responsibility* (Sydney, Australia: St. Louis Press, 1992); Joan Solomon, *Teaching Science, Technology and Society* (Buckingham: Open University Press, 1993); Robert E. Yager, ed., *The Science, Technology, Society Movement* (Washington, D.C.: National Science Teachers Association, 1993), and Yager, ed., *Science/Technology/Society as Reform in Science Education* (Albany: SUNY Press, 1996). Morris H. Shamos's *The Myth of Scientific Literacy* (New Brunswick, N.J.: Rutgers University Press, 1995) is more cautious about the role and value of an STS approach to science education, but it should be considered as an alternative voice. Also interesting is Michael R. Matthews, *Science Teaching: The Role of History and Philosophy of Science* (New York: Routledge, 1994), and one should consider a selection of revised essays by Leonard Waks, *Technology's School: The Challenge to Philosophy*, issued as Supplement 3 of *Research in Philosophy and Technology* (Greenwich, Conn.: JAI Press, 1995).

Supplementary Literature

As a general matter of course throughout this bibliographic essay, I have followed the outlines of the book itself. This means, of course that there are many areas relevant to STS studies, such as medicine, religion and ethics, literature and art, and computers, that I have hardly mentioned if at all. Those wishing to pursue reading in such areas will find a rich body of literature and might begin by consulting relevant chapters in the Durbin and Mitcham and Williams bibliographies mentioned at the beginning of this essay. In this regard, although now out of print, still valuable for the older literature associated with some of these topics is Stephen H. Cutcliffe, Judith A. Mistichelli, and Christine Roysdon, comps., *Technology and Human Values in American Civilization: A Guide to Information Resources* (Detroit: Gale Research, 1980).

Journals

Finally, it is important to say a word about scholarly journals and the article literature. Even in well established fields, let alone those that are still evolving, some of the best scholarship is oftentimes found as journal literature, which makes its way only slowly, and sometimes not at all, into book form. Even though STS as a field of academic inquiry is now reasonably well established after some three decades of scholarship, it is no exception. Therefore, a brief reiteration of some of the key journals within the field is in order. *Social Studies of Science* (Sage) and *Science, Technology, & Human Values* (Sage) are the two central interdisciplinary academic journals in the field, with the former tending to focus on sociological studies of science as knowledge and the latter more broadly including technology and covering historical and political perspectives as well as sociological and anthropological studies. *Perspectives on Science* (MIT Press) also offers an interesting range of constructivist-oriented perspectives. A useful review journal is *Metascience: An International Review Journal for the History, Philosophy and Social Studies of Science* (Blackwell). Three journals—*Technology in Society* (Pergamon), *Research Policy* (Elsevier Science); and *Issues in Science and Technology* (National Academies of Science and Engineering and Institute of Medicine)—deal with policy and management-oriented issues. *Science, Technology, & Society* (Sage) is devoted to international issues of the developing world. For educational issues and curricular materials related to STS, one should read the *Bulletin of Science, Technology & Society* (Sage); the *Science, Technology & Society Curriculum Newsletter* (STS Program, Lehigh University); and the *Teachers' Clearinghouse for Science and Society Education Newsletter* (Trevor Day School, N.Y.C. and Association of Teachers in Independent Schools). Each of these publications frequently includes noncurricular material of general interest as well. *Science as Culture* includes critical studies of technoscience, while *Public Understanding of Science* (Institute of Physics Publishing and The Science Museum, London) deals with issues related to the way scientific and technological information is presented to and received by the general public. Key journals taking a more traditional disciplinary approach to the study of science and technology would include: *Isis* (Johns Hopkins University Press) and *Technology and Culture* (Johns Hopkins University Press) for the history of science and technology respectively; *Philosophy of Science* (University of Chicago Press); *Philosophy and Technology* (Society for Philosophy and Technology, electronic); *Research in Philosophy and Technology* (JAI Press); and *Configurations* (Johns Hopkins), which deals with literary issues. There are of course numerous other journals and newsletters, such as *Knowledge and Society* (JAI Press) that frequently include material of interest to STS scholars and activists alike; the problem, if one can call it that, is that there is an abundance of intellectual riches almost too great for any one person to fully grasp and accommodate. Nonetheless, this richness suggests the importance and vitality of STS as a field of pursuit and study.

Selected Bibliography

The following alphabetical listing of 100 key books is my attempt to identify those works that have had the most influence on STS. To that end I have included a number of works that may appear dated today, but that either have played an important role in the field's development or continue to provide a reference point for contemporary scholars. Thus appear Jacques Ellul's *The Technological Society* (1964), often critiqued for its notion of "technological determinism," and C. P. Snow's more enthusiastic lauding of science, in his somewhat simplistic *Two Cultures* (1964) breakdown. At the same time I have included a number of very recent works that have not yet had time to be accorded status as classics, but that I feel either reflect the current state of STS research or suggest promising future directions. I have also selected titles for this shorter core listing with an eye toward a reasonably broad coverage of important themes and approaches within the field. Finally, I should note that, with one or two key exceptions, I have excluded textbooks, reference works, and other guides from this bibliography.

Barbour, Ian. *Ethics in an Age of Technology*. New York: Harper Collins, 1993.

———. *Religion in the Age of Science*. New York: Harper Collins, 1990.

Barnes, Barry. *About Science*. Oxford: Blackwell, 1985.

Barnes, Barry, David Bloor, and John Henry. *Scientific Knowledge: A Sociological Analysis*. Chicago: University of Chicago Press, 1996.

Beauchamp, Tom L., and James F. Childress. *Principles of Biomedical Ethics*. 4th ed. Oxford: Oxford University Press, 1994.

Biagioli, Mario, ed. *The Science Studies Reader*. New York: Routledge, 1999.

Bijker, Wiebe. *Of Bicycles, Bakelites, and Bulbs: Toward a Theory of Sociological Change*. Cambridge: MIT Press, 1995.

Bijker, Wiebe, and John Law, eds. *Shaping Technology/Building Society*. Cambridge: MIT Press, 1992.

Bijker, Wiebe, Trevor Pinch, and Thomas P. Hughes, eds. *The Social Construction of Technological Systems*. Cambridge: MIT Press, 1987.

Bloor, David. *Knowledge and Social Imagery*. 2d ed. Chicago: University of Chicago Press, 1991.

Blume, Stuart S. *Insight and Industry: On the Dynamics of Technological Change in Medicine*. Cambridge: MIT Press, 1992.

Borgmann, Albert. *Technology and the Character of Contemporary Life*. Chicago: University of Chicago Press, 1987.

Bullard, Robert. *Dumping on Dixie: Race, Class, and Environmental Equity*. Boulder, Colo.: Westview Press, 1990.

Cheek, Dennis. *Thinking Constructively about Science, Technology and Society Education*. Albany: SUNY Press, 1992.

Cole, Stephen. *Making Science: Between Nature and Society*. Cambridge: Harvard University Press, 1992.

Collins, Harry M. *Changing Order: Replication and Induction in Scientific Practice*. Beverly Hills, Calif.: Sage, 1985.

Collins, Harry M., and Trevor Pinch. *The Golem at Large: What You Should Know about Technology*. Cambridge: Cambridge University Press, 1998.

———. *The Golem: What You Should Know about Science*. 2d ed. Cambridge: Cambridge University Press, 1998.

Commoner, Barry. *The Closing Circle*. New York: Alfred A. Knopf, 1971.

Cowan, Ruth Schwartz. *More Work for Mother: The Ironies of Household Technology from the Open Hearth to the Microwave*. New York: Basic Books, 1983.

Cozzens, Susan, and Thomas F. Gieryn, eds. *Theories of Science in Society*. Bloomington: Indiana University Press, 1990.

Cronberg, Tarja, and Knut H. Sorenson, eds. *Similar Concerns, Different Styles? Technology Studies in Western Europe*. Luxembourg: European Commission, 1995.

Cutcliffe, Stephen H., and Carl Mitcham, eds. *Visions of STS: Contextualizing Science, Technology, and Society Studies*. Albany: SUNY Press, 2000.

Cutcliffe, Stephen H., and Robert C. Post, eds. *In Context: History and the History of Technology—Essays in Honor of Melvin Kranzberg*. Bethlehem, Pa.: Lehigh University Press, 1989.

Dickson, David. *The New Politics of Science*. 2d ed. Chicago: University of Chicago Press, 1988.

Dolby, R. G. A. *Uncertain Knowledge: An Image of Science for a Changing World*. Cambridge: Cambridge University Press, 1996.

Durbin, Paul T. *Social Responsibility in Science, Technology, and Medicine*. Bethlehem, Pa.: Lehigh University Press, 1992.

Ellul, Jacques. *The Technological Society*. 1954. Translated by John Wilkinson. New York: Knopf, 1964.

Feenberg, Andrew. *Critical Theory of Technology*. New York: Oxford University Press, 1991.

Florman, Samuel. *The Existential Pleasures of Engineering*. Rev. ed. New York: St. Martin's Press, 1994.

Fujimura, Joan. *Crafting Science: A Sociohistory of the Quest for Genetics of Cancer*. Cambridge: Harvard University Press, 1996.

Fuller, Steve. *Science*. Minneapolis: University of Minnesota Press, 1997.

Galison, Peter. *How Experiments End*. Chicago: University of Chicago Press, 1987.

Gieryn, Thomas F. *Cultural Boundaries of Science: Credibility on the Line*. Chicago: University of Chicago Press, 1998.

Gilbert, G. Nigel, and Michael Mulkay. *Opening Pandora's Box: A Sociological Analysis of Scientists' Discourse*. Cambridge: Cambridge University Press, 1984.

Hamlett, Patrick. *Understanding Technological Politics: A Decision-Making Approach*. Englewood Cliffs, N.J.: Prentice Hall, 1992.

Haraway, Donna. *Primate Visions: Gender, Race, and Nature in the World of Modern Science*. New York: Routledge, 1989.

———. *Simians, Cyborgs, and Women: The Reinvention of Nature*. London: Routledge, 1991.

Harding, Sandra. *Is Science Multicultural? Postcolonialisms, Feminisms, and Epistemologies*. Bloomington: Indiana University Press, 1998.

———. *The Science Question in Feminism*. Ithaca, N.Y.: Cornell University Press, 1986.

Hays, Samuel P. *Beauty, Health and Permanence: Environmental Politics in the United States, 1955-1985*. Cambridge: Cambridge University Press, 1987.

Hess, David J. *Science and Technology in a Multicultural World: The Politics of Facts and Artifacts*. New York: Columbia University Press, 1995.

———. *Science Studies: An Advanced Introduction*. New York: New York University Press, 1997.

Hughes, Thomas P. *American Genesis: A Century of Invention and Technological Enthusiasm, 1870–1970*. New York: Viking, 1989.

Hull, David. *Science as a Process: An Evolutionary Account of the Social and Conceptual Development of Science*. Chicago: University of Chicago Press, 1988.

Ihde, Don. *Technology and the Lifeworld: From Garden to Earth*. Bloomington: Indiana University Press, 1990.

Illich, Ivan. *Tools for Conviviality*. New York: Harper and Row, 1973.

Jasanoff, Sheila. *The Fifth Branch: Science Advisors as Policy Makers*. Cambridge: Harvard University Press, 1990.

———. *Science at the Bar: Law, Science, and Technology in America*. Cambridge: Harvard University Press, 1995.

Jasanoff, Sheila, Gerald E. Markle, James C. Petersen, and Trevor Pinch, eds. *Handbook of Science and Technology Studies*. Thousand Oaks, Calif.: Sage, 1995.

Johnson, Deborah. *Computer Ethics*. Englewood Cliffs, N.J.: Prentice Hall, 1985.

Keller, Evelyn Fox. *Reflections on Gender and Science*. New Haven: Yale University Press, 1985.

Kitcher, Philip. *Advancement of Science: Science without Legend, Objectivity without Illusions*. Oxford: Oxford University Press, 1993.

Knorr-Cetina, Karin. *Epistemic Cultures: How the Sciences Make Knowledge*. Cambridge: Harvard University Press, 1998.

———. *The Manufacture of Knowledge: An Essay on the Constructivist and Contextual Nature of Science*. New York: Pergamon Press, 1981.

Kuhn, Thomas. *The Structure of Scientific Revolutions*. 2d ed. Chicago: University of Chicago Press, 1970.

Latour, Bruno. *Pandora's Hope: Essays on the Reality of Science Studies*. Cambridge: Harvard University Press, 1999.

———. *Science in Action: How to Follow Scientists and Engineers through Society*. Cambridge: Harvard University Press, 1987.

Latour, Bruno, and Steve Woolgar. *Laboratory Life: The Social Construction of Scientific Facts*. 2d ed. Princeton: Princeton University Press, 1986.

Longino, Helen. *Science as Social Knowledge: Values and Objectivity*. Princeton: Princeton University Press, 1990.

Lynch, Michael. *Scientific Practice and Ordinary Action: Ethnomethodology and Social Studies of Science*. Cambridge: Cambridge University Press, 1993.

MacKenzie, Donald. *Inventing Accuracy: A Historical Sociology of Nuclear Missile Guidance*. Cambridge: MIT Press, 1990.

———. *Knowing Machines: Essays on Technological Change*. Cambridge: MIT Press, 1996.

Merchant, Carolyn. *The Death of Nature: Women, Ecology and the Scientific Revolution*. New York: Harper and Row, 1980.

Merton, Robert K. *The Sociology of Science*. Chicago: University of Chicago Press, 1973.

Mitcham, Carl. *Thinking through Technology: The Path between Engineering and Philosophy*. Chicago: University of Chicago Press, 1994.

Morone, Joseph, and Edward J. Woodhouse. *Averting Catastrophe: Strategies for Regulating Risky Technologies*. Berkeley: University of California Press, 1986.

Mulkay, Michael. *Sociology of Science: A Sociological Pilgrimage*. Bloomington: Indiana University Press, 1991.

Mumford, Lewis. *The Myth of the Machine*. Vol. 1: *Technics and Human Development*; Vol. 2: *Pentagon of Power*. New York: Harcourt Brace, Jovanovich, 1967, 1970.

———. *Technics and Civilization*. New York: Harcourt, Brace and Company, 1934.

Nelkin, Dorothy, ed. *Controversy: Politics of Technical Decisions*. 3d ed. Newbury Park, Calif.: Sage, 1992.

Pickering, Andrew. *Constructing Quarks: A Sociological History of Particle Physics*. Chicago: University of Chicago Press, 1984.

———. *The Mangle of Practice: Time, Agency, and Science*. Chicago: University of Chicago Press, 1995.

———. ed. *Science as Practice and Culture*. Chicago: University of Chicago Press, 1992.

Postman, Neil. *Technopoly: The Surrender of Culture to Technology*. New York: Knopf, 1992.

Proctor, Robert, *Cancer Wars*. Cambridge: Harvard University Press, 1995

Rouse, Joseph. *Engaging Science: How to Understand Its Practices Philosophically*. Ithaca, N.Y.: Cornell University Press, 1996.

Ruse, Michael. *Mystery of Mysteries: Is Evolution a Social Construction?* Cambridge: Harvard University Press, 1999.

Sarewitz, Daniel. *Frontiers of Illusion: Science, Technology and the Politics of Progress*. Philadelphia: Temple University Press, 1996.

Schiebinger, Londa. *Has Feminism Changed Science?* Cambridge: Harvard University Press, 1999.

Schumacher, E. F. *Small Is Beautiful: Economics as if People Mattered.* New York: Harper and Row, 1973.

Sclove, Richard E. *Democracy and Technology.* New York: Guilford, 1995.

Shapin, Steven, and Simon Schaffer. *Leviathan and the Air-Pump: Hobbes, Boyle, and the Experimental Life.* Princeton: Princeton University Press, 1985.

Simon, Julian, ed. *The State of Humanity.* Cambridge: Blackwell, 1995.

Smith, Bruce L. R. *American Science Policy Since World War II.* Washington, D.C.: Brookings Institute, 1990.

Snow, C. P. *The Two Cultures: And a Second Look.* Cambridge: Cambridge University Press, 1964.

Spiegel-Rosing, Ina, and Derek de Solla Price. *Science, Technology, and Society: A Cross-Disciplinary Perspective.* London: Sage, 1977.

Staudenmaier, John. *Technology's Storytellers: Reweaving the Human Fabric.* Cambridge: MIT Press, 1985.

Strathern, Marilyn. *Reproducing the Future: Anthropology, Kinship and the New Reproductive Technologies.* New York: Routledge, 1992.

Traweek, Sharon. *Beamtimes and Lifetimes: The World of High Energy Physicists.* Cambridge: Harvard University Press, 1988.

Unger, Stephen. *Controlling Technology: Ethics and the Responsible Engineer.* 2d ed. New York: Wiley, 1993.

Volti, Rudi. *Society and Technological Change.* 3d ed. New York: St. Martin's Press, 1995.

Wajcman, Judy. *Feminism Confronts Technology.* University Park: Penn State University Press, 1991.

Webster, Andrew. *Science, Technology, and Society.* New Brunswick, N.J.: Rutgers University Press, 1991.

Wenk, Edward. *Making Waves: Engineering, Politics and the Social Management of Technology.* Urbana: University of Illinois Press, 1995.

Winner, Langdon. *Autonomous Technology: Technics Out-of-Control as a Theme in Political Thought.* Cambridge: MIT Press, 1977.

———. *The Whale and the Reactor: A Search for Limits in an Age of High Technology.* Chicago: University of Chicago Press, 1986.

Woolgar, Steve. *Science: The Very Idea.* London: Tavistock, 1988.

Yager, Robert E., ed. *Science/Technology/Society as Reform in Science Education.* Albany: SUNY Press, 1996.

Ziman, John. *An Introduction to Science Studies: The Philosophical and Social Aspects of Science and Technology.* Cambridge: Cambridge University Press, 1984.

Index

About the Author

Stephen H. Cutcliffe is a historian of technology and Director of the Science, Technology, and Society program at Lehigh University, where he also edits the *Science, Technology & Science Curriculum Newsletter*. He has served as President of the National Association of Science, Technology, and Society and co-edited several volumes of essays on STS and history of technology topics. Among them are: *In Context: History and the History of Technology—Essays in Honor of Melvin Kranzberg* (Lehigh University Press, 1989); *Technology and American History* (University of Chicago Press, 1997); and *Visions of STS: Contextualizing Science, Technology, and Society Studies* (SUNY Press, 2000).